From Neutron Resonances

to

Random Matrices

A collection of review articles by

Frank W. K. Firk

Professor Emeritus of Physics, Yale University

2019

II

Contents

The page numbers of the original published articles are listed.

IV

1. Introduction

Nuclear Physics began with Chadwick's discovery of the neutron in early 1932 (extensive references are given in the seven articles). Chadwick concluded from an analysis of his data that the neutron mass is slightly greater than the proton mass. In the same year, Cockcroft and Walton carried out the first proton-induced nuclear reaction using a high-voltage generator that produced protons with energies up to 600 keV.

Heisenberg introduced a theoretical model of the neutron-proton interaction in which he treated the neutron and proton as two charge states of a single object, the *nucleon*. Heisenberg's model included space, spin and charge exchange terms in the nuclear Hamiltonian. Fermi introduced a model of β-decay in which a neutron decays into a proton, a spontaneously generated electron and an anti-neutrino. In this model, a neutron is a fundamental particle and not a proton-electron combination as assumed in the original Heisenberg model. Wigner showed that the nucleon-nucleon force is characterized by a short-range attractive potential and a repulsive potential (core) between the nucleons when they are in very close proximity. Breit et al. analyzed data on proton-proton scattering and showed that the nuclear force is the same between neutron-proton and proton-proton pairs, after correcting for the Coulomb interaction. This result led to the hypothesis of *charge independence* for nuclear forces Yukawa proposed a novel mechanism to describe the neutron-proton interaction. In his model, the interaction between nucleons is mediated by a heavy particle, the quantum of the nuclear force field. Based on the limited data available at the time, he estimated the mass of the field quantum (a boson) to be 200 electron masses. Yukawa's model evolved into contemporary Particle Physics theory.

The discovery of a quadrupole moment of the deuteron led to the introduction of a non-central component of the nuclear force. This component, called a tensor force, mixes states with the same total angular momentum quantum number but with different orbital and spin quantum numbers.

The discovery of neutron-induced fission of certain heavy nuclei changed global politics and economics forever. Almost immediately, Bohr and Wheeler published a theoretical model of the fission process. The field of Reactor Physics was born.

In 1950, a spin-orbit component of the nucleon interaction was introduced following the success of the shell model of nuclear structure, pioneered by Mayer and Jensen. The magnitude and sign of the spin-orbit part of the nucleon-nucleus potential were deduced by Adair in studies of the polarization of fast neutrons scattered elastically from medium mass nuclei. Bohr and Mottelson described the collective motions of excited nuclei geometrically, and later Arima and

Iachello developed the geometric model using algebraic methods (the interacting boson approximation).

The first studies of low-energy neutron induced nuclear reactions showed characteristic resonant states at excitation energies of more than 5 MeV in a typical nucleus. The observation of these states led Niels Bohr to introduce the compound nucleus model in which an incident neutron enters the potential well of the target nucleus and is trapped. The compound state does not decay until sufficient energy resides in one of the nucleons for it to escape, or the state decays by emitting a sequence of gamma rays. This is a statistical process and therefore it can take a long time (on a nuclear time scale) for the compound system to give up its excitation energy. Breit and Wigner introduced a theory of the resonant form. Kapur and Peierls and Wigner and Eisenbud generalized the theory. The Wigner-Eisenbud version has been used to analyze almost all nuclear reaction data since that time.

In the period between 1955 and 1960 the observed level spacing distribution of neutron resonances of the same spin and parity in heavy nuclei verified the surmised Wigner distribution. This was the beginning of Random Matrix Theory (RMT) as it is known today. The impact of RMT in fields as diverse as Number Theory, Condensed Matter Physics and String Theory has far exceeded the expectations of pioneers in the field.

4

The study of nuclear photo-disintegration began when Chadwick and Goldhaber used gamma rays from a radioactive source as a means of photo-disintegrating the deuteron; they thereby obtained an improved value for the mass of the neutron. In 1948, Baldwin and Klaiber reported the measurement of the cross section for the nuclear absorption of gamma rays with energies in the region of 20 MeV. Their results were interpreted by Goldhaber and Teller in terms of collective oscillations of all protons relative to all neutrons in a given nucleus. The so-called giant electric dipole resonances are a feature of all nuclei. In light and medium mass nuclei discrete resonances are observed[)]; the resonances are interpreted in terms of photon absorption into the underlying shell structure of the nuclei.

This collection of articles covers theoretical and experimental aspects of neutron and photoneutron spectroscopy during the half-century, beginning in the early 1930's. This was the golden age of classical Nuclear Physics. The topics reviewed include the methods used to study total neutron cross sections, developments in the detection of neutrons, time-of-flight methods, neutron polarization measurements and analyses, studies of photonuclear reactions, and the key connection between the spacing distribution of resonances in heavy nuclei, random matrix theory and number theory.

TOTAL NEUTRON CROSS SECTION MEASUREMENTS

F. W. K. FIRK
YALE UNIVERSITY
NEW HAVEN, CONNECTICUT

E. MELKONIAN
COLUMBIA UNIVERSITY
NEW YORK, NEW YORK

6

I. INTRODUCTION

A. General Considerations

During the past twenty years, numerous papers have reported measurements of total neutron cross sections for incident neutron energies up to 50 keV [see, for example, the compilations of data by Hughes and Schwartz (1958) and Stehn *et al.* (1965)]. The reasons for the popularity of this branch of low energy nuclear physics are twofold, namely: the relative simplicity of the experiments discussed in this section and the significance of the results in both reactor physics and theoretical nuclear physics, discussed in Section V.

A total neutron cross section is defined as

$$\frac{\text{number of events of all types per unit time per nucleus}}{\text{number of incident neutrons per unit area per unit time}}$$

or, qualitatively, as the effective area which a target nucleus presents to an incoming neutron. The outstanding feature of low energy total neutron cross sections, measured as a function of incident neutron energy, is the appearance of sharp resonances which correspond to discrete energy levels in the compound state formed by the target nucleus plus neutron. (We are concerned here only with nuclear effects and not with atomic, molecular, magnetic, or crystalline effects observed at very low neutron energies.)

A total neutron cross section $\sigma_T(E)$ is determined by measuring the transmission $T(E)$ of monoergic neutrons of energy E through an element of uniform thickness. This is related to the cross section by the equation

$$T(E) = \exp\{-n\sigma_T(E)\} \tag{1}$$

where n is the number of nuclei per square centimeter of the element normal to the incident neutron beam. The detector of the transmitted neutrons must subtend a small solid angle at the sample so that elastically scattered neutrons are not detected. Furthermore, the incident neutron beam must be suitably collimated at the sample position in order that the number of nuclei per square centimeter of sample, normal to the beam, is well defined. The experiment involves only the determination of the ratio of counting rate in a neutron detector with and without an absorber in the neutron beam and a measurement of the neutron energy. No absolute measurement of the incident neutron flux is required. Corrections for multiple scattering of neutrons within the sample, or for the attenuation of reaction γ-rays by the sample (as required in partial cross-section measurements) are unnecessary since a transmission measurement is not concerned with the details of a nuclear reaction; it is simply concerned with the fact that a nuclear encounter has taken place and has thereby prevented the neutron from reaching the detector.

Although fundamentally straightforward, there are several difficulties associated with measuring the true total neutron cross section as given by Eq. (1). First, the neutron spectrometer used to measure the neutron energy E' has a finite resolving power so that the observed transmission $T_{obs}(E)$ is given by

$$T_{obs}(E) = \int_{E_1}^{E_2} R(E' - E)T(E')\, dE' \tag{2}$$

where $R(E' - E)$ is the instrumental resolution function such that $R(E' - E) = 0$ for $E_1 > E' > E_2$. Second, the target nucleus is not at rest in the laboratory system but has a distribution of velocities which corresponds to the thermal motion of atoms in the lattice of the sample material. This so-called Doppler effect can drastically distort the shapes .of narrow resonances which appear in the total neutron cross section. The above two effects are treated in detail in Section III,A.

B. Theoretical Considerations

The form of the cross section for a reaction proceeding through an isolated resonance was first given by Breit and Wigner (1936). The basis for their initial work was not well founded, however, because time-dependent perturbation theory had been used to describe strong nuclear interactions. Nevertheless, the shape of the cross section agreed well with observation due to the fact that the shape only depends upon the formation of a long-lived intermediate state. Such a compound state is a consequence of the strength of nuclear forces.

A rigorous theory of nuclear reactions was introduced by Kapur and Peierls (1938) which does not depend upon the approximations of the perturbation method and, furthermore, does not depend upon particular reaction mechanisms such as compound nucleus formation. An important feature of their theory (and many later developments) is the occurrence of a complete set of formal states, defined in a volume of nuclear dimensions, which result from imposing certain boundary conditions at the surface of the volume. The theory is particularly well suited to a description of compound nucleus states which can be identified with the formal states. The Kapur–Peierls boundary conditions are energy-dependent and complex so that most of the parameters which occur are also energy-dependent. Their theory, therefore, has not become fashionable in the interpretation of low energy nuclear reactions. Wigner and Eisenbud (1947) introduced different boundary conditions that lead to parameters that have a clearer physical significance.

Breit (1940, 1946) attempted to remove from the theory those parameters, such as interaction radii, whose values cannot affect the cross sections. More recently, Humblet and Rosenfeld (1961) have introduced a theory that places emphasis on the behavior of the wave functions in the external region and on a correct description of the cross sections themselves. Again, it is not straightforward to interpret measured resonance parameters due to the appearance of complex phase shifts in the "widths" of the decaying states. We shall therefore use the Wigner–Eisenbud (1947) theory bearing in mind, however, that all the theories since Kapur and Peierls are formally correct. The following outline shows how the significant quantities which occur in discussions of slow neutron interactions appear in the formal theory [for a detailed discussion see Lane and Thomas (1958)].

Consider the problem of the elastic scattering of an s-wave neutron (assumed spinless) by a square potential well (Vogt, 1962). The radial part of the Schrödinger equation is then

$$-(\hbar^2/2m)(d^2\phi/dr^2) + V\phi = E\phi \tag{3}$$

where

$$V = V_1 \quad \text{for} \quad r \lessgtr a; \quad V = 0 \quad \text{for} \quad r > a$$

The mass of the particle is m and the radius of the well is a.

The solutions of Eq. (3) are

$$\phi = A \sin Kr \quad \text{for} \quad r \lessgtr a \tag{4}$$

and

$$\phi = B(e^{-ikr} - Ue^{ikr}) \quad \text{for} \quad r > a \tag{5}$$

where A is an arbitrary constant, K and k are the wave numbers inside and outside the well, respectively, and B is a normalizing factor chosen so that the incoming wave has unit flux

$$B = (4\pi v)^{-1/2} \tag{6}$$

where v is the neutron velocity. Here, U is termed the collision function which may also be defined in terms of a phase shift

$$U = e^{2i\delta} \tag{7}$$

The scattering cross section σ_{nn} can then be written

$$\sigma_{nn} = (\pi/k^2)|1 - U|^2 = (4\pi/k^2)\sin^2\delta \tag{8}$$

The wave function ϕ which describes the scattering resonance at low energies is almost a standing wave. The term "almost" applies because the neutron is unbound and eventually leaves the well after a time interval $\sim 10^{-14}$ sec.

The formal theory is concerned with constructing a complete set of standing (stationary) waves X_λ of which one closely resembles the true wave function ϕ.

In expanding the wave function ϕ in terms of the standing waves X_λ, one term may dominate the expansion and this term then corresponds to the standing wave associated with the resonance.

The stationary states X_λ are solutions of the equation

$$-(\hbar^2/2m)(d^2X_\lambda/dr^2) + VX_\lambda = E_\lambda X_\lambda \tag{9}$$

subject to the boundary condition at the radius a

$$[r(dX_\lambda/dr) = bX_\lambda]_{r=a} \tag{10}$$

where b is an arbitrary number. The X_λ form a complete orthogonal set so that the actual wave function may be expanded, thus,

$$\phi = \sum_\lambda A_\lambda X_\lambda \tag{11}$$

where

$$A_\lambda = \int_0^a X_\lambda \phi \, dr \tag{12}$$

Multiplying Eq. (3) by X_λ and Eq. (9) by ϕ, subtracting and integrating by parts twice, and using Eq. (10), we find

$$(\hbar^2/2ma)\{a\phi'(a) - b\phi(a)\}X_\lambda(a) = \int_0^a (E_\lambda - E)X_\lambda \phi \, dr \tag{13}$$

where

$$\phi' = d\phi/dr$$

Using Eqs. (12) and (13) gives

$$A_\lambda = (E_\lambda - E)^{-1}(\hbar^2/2ma)X_\lambda(a)\{a\phi'(a) - b\phi(a)\} \tag{14}$$

so that

$$\phi = \sum_\lambda (E_\lambda - E)^{-1}(\hbar^2/2ma)X_\lambda^2(a)\{a\phi'(a) - b\phi(a)\} \tag{15}$$

$$= R\{a\phi'(a) - b\phi(a)\} \tag{16}$$

where

$$R = \sum_\lambda \gamma_\lambda^2/(E_\lambda - E) \tag{17}$$

and

$$\gamma_\lambda{}^2 = (\hbar^2/2ma)X_\lambda{}^2(a) \tag{18}$$

The R-function, as defined by Eq. (17), is one of the principal quantities appearing in the formal theory.

The collision function U can be obtained in terms of R by considering the logarithmic derivative ϕ'/ϕ at $r = a$, giving

$$\phi'(a)/\phi(a) = (1 + bR)/aR \tag{19}$$

Equation (5) is then rewritten as follows:

$$\phi = I - UO \tag{20}$$

where

$$I = (4\pi v)^{-1/2}e^{-ikr} \qquad \text{and} \qquad O = I^*$$

the incoming and outgoing waves, respectively. Substituting these values of $\phi(a)$ and $\phi'(a)$ in Eq. (19), the collision function becomes

$$U = e^{-2ika}\left(\frac{1 + bR + ikaR}{1 + bR - ikaR}\right) \tag{21}$$

If there is only a single resonance centered at an energy E_0, say, the R-function is dominated by one term

$$R_0 \approx \gamma_0{}^2/(E_0 - E) \tag{22}$$

and the cross section close to the resonance becomes

$$\sigma_{nn}^0 = \frac{\pi}{k^2}\left| 2\sin kae^{ika} - \frac{\Gamma_0}{(E_0 - E + \Delta_0) - \frac{1}{2}\Gamma_0}\right|^2 \tag{23}$$

where $\Gamma_0 = 2ka\gamma_0{}^2$ is the width of the level and $\Delta_0 = b\gamma_0{}^2$ is the level shift which shifts the maximum of the resonance from E_0 to $E_0 + \Delta_0$.

The simplest boundary condition is such that $b = 0$; the real resonance states have zero logarithmic derivative at $r = a$ so that when $b = 0$, the standing wave resembles the true scattering state as closely as possible. Furthermore, when $b = 0$, the level shift is zero so that the eigenvalue of the standing wave problem coincides with the energy of the peak cross section.

The above theory may be extended to the general case of nuclear reactions without changing the basic principles involved in generating the R-function. In practice, however, there are frequently too many details to handle in a satisfactory way. These difficulties are due to the fact that, in general, an excited state of a nucleus can lose energy in many different ways; for example, by emitting neutrons, protons, γ-radiation, etc. These different modes of decay are referred to as channels. The simple R-function obtained above for the case of elastic scattering must be replaced by an R-matrix in which the value of the wave function for a channel, at the nuclear surface, is related to the value of the derivative of the wave function for all the channels, also at the surface. The rows and columns of the R-matrix refer to the different channels, and the expressions for I, O, U, and b also become matrices. Equation (21) still applies when written in a matrix notation.

In its most general form, the total neutron cross section $\sigma_{n,T}(J)$ may be written [see, for example, Vogt (1959)]

$$\sigma_{n,T}(J) = (2\pi/k^2)g(J) \sum_{s=|I-i|}^{I+i} \sum_{l=|J-s|}^{J+s} [1 - \text{Re } U_{nsl,nsl}(J)] \qquad (24)$$

where k is the wave number of the relative motion of the incident neutron and the target nucleus and $g(J)$ is a statistical weighting factor where

$$g(J) = \frac{2J+1}{(2i+1)(2I+1)}$$

I is the ground state spin of the target nucleus;

i is the intrinsic spin of the incident neutron ($i = \frac{1}{2}$);

J is the total angular momentum of the compound state ($\mathbf{J} = \mathbf{l} + \mathbf{s}$);

s is the channel spin ($\mathbf{s} = \mathbf{I} + \mathbf{i}$);

l is the orbital angular momentum of the relative motion of the incident neutron and target nucleus; and

Re $U_{nsl,nsl}(J)$ is the real part of the collision function diagonal element $U_{nsl,nsl}$ defined as in Eq. (21) when written in an appropriate matrix (many-channel) form.

For neutron energies up to 50 keV, the most probable processes that occur are elastic scattering of the incident neutron and radiative capture. The case in which the compound state undergoes fission is treated in Chapter V and will not be considered here.

The major problem in the practical application of the formal theory is to develop a suitable form for the collision function U. Thomas (1955) obtained particularly useful expressions for U and his approach will be used here.

At neutron energies below 50 keV nuclear reactions are predominantly s-wave ($l = 0$) since the penetration factors for incident neutrons of higher orbital angular momenta are small. Exceptions occur for nuclei with mass numbers $90 \gtrsim A \gtrsim 110$ and $A \gtrsim 230$. In these cases, p-wave ($l = 1$) interactions have been observed at neutron energies below 1 keV.

The following formulas are developed for s-wave interactions: the extension of the theory to include higher l-values is tedious but straightforward [see Lane and Thomas (1958)].

If the expression for the total cross section is restricted to include a single component of total angular momentum J, Eq. (24) becomes

$$\sigma_{n,T} = (2\pi/k^2)g(1 - \text{Re } U_{nn}) \qquad (l = 0) \tag{25}$$

Thomas (1955) showed that for elastic scattering and radiative capture of s-wave neutrons, the diagonal elements U_{nn} of the collision function can be written

$$U_{nn} = \exp\{-2ika\}(1 + ikaR_{nn})/(1 - ikaR_{nn}) \tag{26}$$

where the reduced R-function R_{nn} is

$$R_{nn} = \sum_{\lambda} \gamma_{\lambda n}^2/(E_\lambda - E - \tfrac{1}{2}i\Gamma_{\lambda\gamma}) \tag{27}$$

in which

E_λ is the resonance energy of level λ;
$\gamma_{\lambda n}^2$ is the reduced neutron width ($\gamma_{\lambda n}^2 = \Gamma_{\lambda n}/2ka$);
$\Gamma_{\lambda n}$ is the neutron width of level λ;
$\Gamma_{\lambda\gamma}$ is the total radiation width of level λ; a is the nuclear radius; and
k is the wave number of the incident neutron.*
The level shift is assumed to be zero.

Equation (27) is a good approximation provided that all the partial radiation widths [the eliminated channels in Thomas (1955)] are much less than the level spacing and, furthermore, that their amplitudes are random in sign. The collision function U_{nn} given by Eq. (26) is valid irrespective of whether the total widths Γ_λ are greater or smaller than the level spacings.

If there is only one level in the sum over λ of Eq. (27), then the total cross section becomes

$$\sigma_{n,T} = \frac{\sigma_0}{1 + x^2} + \sigma_0 \tan(2ka) \cdot \frac{x}{1 + x^2} + \frac{4\pi}{k^2} g \sin^2(ka) \tag{28}$$

where

$$\sigma_0 = (4\pi/k^2)g(\Gamma_n/\Gamma) \cos(2ka)$$

is the peak total cross section

---

* In terms of the laboratory energy E_L of the incident neutron, the wave number k is

$$k = \frac{2\pi}{\lambda} = \left(\frac{M_r}{M_r + 1}\right)[2.1968 \times 10^9 \times (E_L(\text{eV}))^{1/2}] \quad \text{cm}^{-1}$$

where λ is the de Broglie wavelength, M_r the mass of the target nucleus.

$$x = (2/\Gamma)(E - E_\lambda), \qquad \text{and} \qquad \Gamma = \Gamma_n + \Gamma_\gamma$$

Equation (28) is the Breit–Wigner single level form for the total cross section in the neighborhood of a resonance of total angular momentum J (Breit and Wigner, 1936). It consists of three parts:

(i) a resonance term: $\sigma_{\text{res}} = \sigma_0/(1 + x^2)$
(ii) an interference term: $\sigma_{\text{int}} = \sigma_0 \tan(2ka) \cdot [x/(1 + x^2)]$
(iii) a term representing part of the potential or hard sphere scattering.

For nuclei with $I \neq 0$, resonances of both spin states $(I \pm \frac{1}{2})$ are present, in which case, the total cross section is obtained by adding the contributions as indicated in Eq. (24).

If the contribution from the opposite spin state is added to the last term of Eq. (28), then the potential scattering cross section σ_{pot} is obtained

$$\sigma_{\text{pot}} = (4\pi/k^2) \sin^2(ka) \approx 4\pi a^2 \qquad \text{for} \quad ka \ll 1 \qquad (29)$$

This is the cross section of an impenetrable sphere of radius a.

The cumulative effect of distant levels on the cross section may be found by writing Eq. (27) as follows:

$$R_{nn} = \frac{\gamma_{\lambda n}^2}{E_\lambda - E - \frac{1}{2}i\Gamma_{\lambda\gamma}} + R_{nn}^\infty \qquad (30)$$

Here, R_{nn}^∞ is the contribution from all other levels of the same spin and parity as the isolated level under consideration.

If Eq. (30) is written in the form

$$R_{nn} = A + iB$$

then the total cross section becomes

$$\sigma_{n,T} = \frac{2\pi}{k^2} g \left\{ \frac{\rho^2(A^2 + B^2)(1 + \cos\phi) + 2\rho(B - A \sin\phi) + 1 - \cos\phi}{1 + 2\rho B + \rho^2(A^2 + B^2)} \right\} \qquad (31)$$

where $\rho = ka$ and $\phi = 2ka$.

If R_{nn}^∞ is assumed to be a real constant (no absorption effects included), then the effect of adding this contribution from distant levels may be clearly seen by considering the approximations

$$2(E_\lambda - E) \gg \Gamma \qquad \text{and} \qquad ka \ll 1$$

in which case, we obtain

$$\sigma_{n,T} \approx \frac{\sigma_0 \Gamma^2}{4(E_\lambda - E)^2} - \frac{\sigma_0 \Gamma ka(1 - R_{nn}^\infty)}{(E_\lambda - E)} + 4\pi g[a(1 - R_{nn}^\infty)]^2 \qquad (32)$$

In this low energy approximation, the effect of distant levels is to modify the nuclear radius. Using the notation of Feshbach *et al.* (1954), we have

$$R' = a(1 - R_{nn}^{\infty}) \tag{33}$$

More general results are given by Lane and Thomas (1958) in which modifications to the level shift and partial widths are discussed.

The importance of including effects of R_{nn}^{∞} is demonstrated in Sections III,A and III,C.

II. EXPERIMENTAL TECHNIQUES

A. Typical "Good-Geometry" Arrangements

In the previous section, it was pointed out that to measure total cross sections by the transmission method, the neutron detector must not detect neutrons elastically scattered from the sample. Also, the incident neutron beam must be well collimated at the sample position in order that the number of nuclei per square centimeter of sample is accurately defined. A "good geometry" arrangement is therefore used, whereby the sample subtends a small solid angle at the source and at the detector.

The four principal sources of neutrons used for measurements of total cross sections in the energy range of interest are: (i) neutron choppers, (ii) pulsed electron linear accelerators and pulsed cyclotrons, (iii) crystal spectrometers, and (iv) pulsed and dc Van de Graaff machines. Each of these devices has characteristics (treated in detail in other chapters) which lead to different experimental arrangements in transmission measurements.

The accuracy of total cross section measurements is frequently limited by uncertainties in the magnitude of the background. The following treatment gives a typical method for determining the background in a time-of-flight spectrometer. In such spectrometers, the background can generally be resolved into a time-constant component T and an energy (or time-of-flight) dependent component $B(t)$. At time τ, the observed neutron spectrum is simply the sum of T, $B(\tau)$, and $C(\tau)$ where $C(\tau)$ is the true spectrum. The energy-dependent background $B(\tau)$ arises from neutrons with an apparent flight time τ which originate from source neutrons with high energies of average flight time $\bar{\tau}$ ($< \tau$). These high energy neutrons are scattered into the neutron detector and its shielding, and are finally detected at a later time τ after being moderated in the surrounding materials.

If a sample, that has a strong resonance at time τ, is placed in the beam, then all the true spectrum neutrons $C(\tau)$ can be prevented from reaching the detector thus leaving only the background components. The observed sample-in spectrum $IN_{obs}(\tau)$ is then given by

$$IN_{obs}(\tau) = T + B(\tau) \exp\{-n\sigma_{eff}(\bar{\tau})\} \qquad (34)$$

Thus, $B(\tau)$ is determined provided $\sigma_{eff}(\bar{\tau})$ is known. The effective cross section $\sigma_{eff}(\bar{\tau})$ for the removal of fast neutrons can be obtained by making measurements with successively "blacker" samples. For the Harwell 40-MeV electron linear accelerator, $\sigma_{eff}(\bar{\tau})$ corresponds to the total cross section of a particular sample averaged over the energy range 500–1000 keV.

The "time-constant" background T may fluctuate during a measurement due to changes in the general background level (caused, for example, by switching on or off a nearby reactor or accelerator). It is then usual to monitor T by means of a timing channel (background gate) which opens at the end of a machine cycle. In most spectrometers it is found that the energy-dependent background $B(l)$ is negligible at very long flight times .so that T may be readily determined.

Another important point in measuring total cross sections is the need for good stability in the efficiency of the neutron detection system and beam monitoring sytsem. Measurements often take many days to complete so that stability is of prime importance in relating the sample-in and sample-out results. The effects of drifts in the efficiency of the system may be greatly reduced by using "sample changers" which automatically change from the sample-in to the sample-out condition at short intervals of about 10 min duration. The use of two or more independent beam monitors is also advisable when cross section data with accuracies of better than 5% are required.

Typical examples of high quality total cross section data currently available are shown in Figs. 1–3. "High quality" is used here in the sense of good energy resolution coupled with good statistical accuracy. Some measurements have the additional advantage of the use of small quantities of samples (in particular separated isotopes).

The major installations carrying out total cross section measurements have already been referred to in Chapter I.

B. Neutron Detectors

A variety of neutron detectors have been used for transmission measurements from ~ 1 eV to 50 keV. The factors which are desirable in any particular detector may be listed as follows:

16

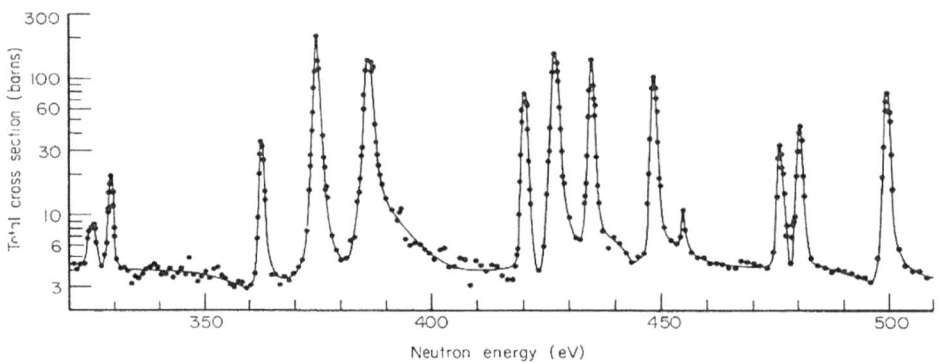

Fig. 1

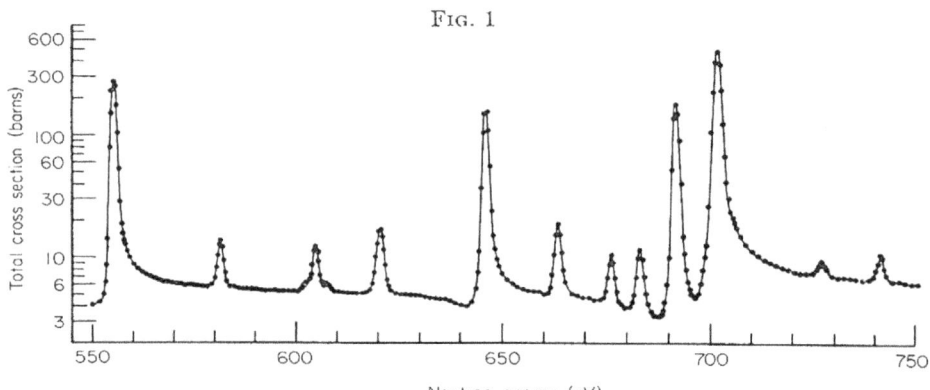

Fig. 2

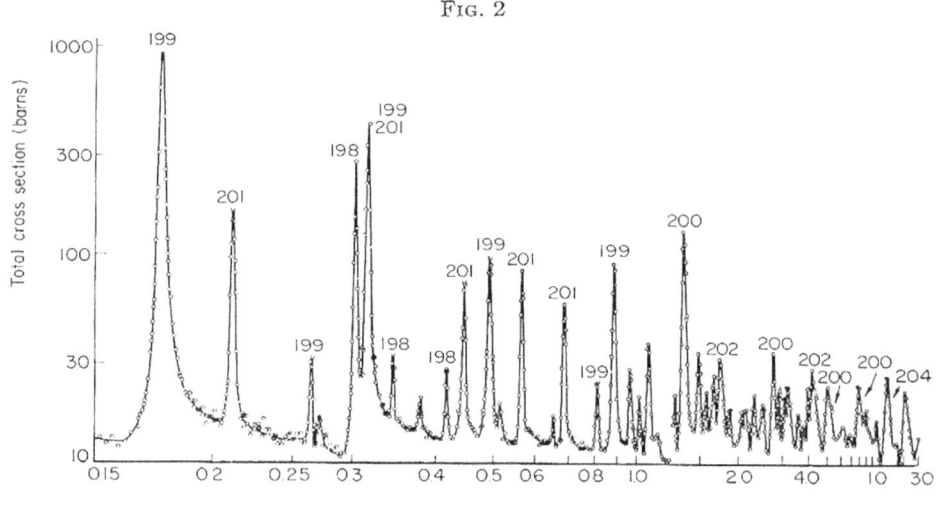

Fig. 3

(i) High detection efficiency which is a smooth function of neutron energy.
(ii) Timing resolution compatible with the duration of the neutron burst for time-of-flight applications.
(iii) Low sensitivity to fast neutron and γ-ray backgrounds.
(iv) Ease of production in various shapes and sizes.
(v) Simplicity of the associated electronics.

In most cases, the reactions $^{10}B(n,\alpha\gamma)^{7}Li + 2.3$ MeV or $^{6}Li(n,\alpha)T + 4.8$ MeV are used. The thermal capture cross sections for these reactions are large: ~4000 b and ~1000 b, respectively. They exhibit a smooth $1/V$ dependence in the energy range of interest. A number of other reactions have been used, e.g., $Sm(n, \gamma)$ and $Ag(n, \gamma)$, but these have not found widespread application due to low average capture cross sections and pronounced resonance structure.

Although the $^{10}B(n,\alpha\gamma)$ and $^{6}Li(n,\alpha)$ reactions are common to most slow neutron detectors, the detection mechanism varies considerably. For example, the ions created in a $^{10}BF_3$ gas counter due to the passage of an α-particle, may be collected and a measurable current obtained. Alternatively, the 480 keV γ-ray which results more than 90% of the time when ^{10}B captures a slow neutron, may be observed with a NaI(Tl) scintillation counter (Rae and Bowey, 1953). Recently, the scintillations produced by the heavily ionizing reaction products in boron- and lithium-loaded glass scintillators have been successfully observed (Voitovetskii et al., 1959; Bollinger et al., 1959; Firk et al., 1961; and Bollinger et al., 1962).

The more important characteristics of each type of detector are listed in Table I. These data are mostly taken from a compilation by Brooks (1961). It is evident from Table I that a number of detectors combine reasonable efficiencies with the demanding requirement of fast time resolution.

A frequent problem is that of the sensitivity of neutron detectors to background γ-rays. This has been discussed by Brooks (1961) who made

---

Fig. 1. The observed total neutron cross section of iodine (Garg et al., 1965). This measurement was made using the Columbia University pulsed synchrocyclotron with a time-of-flight resolution of 0.5 nsec m⁻¹.

Fig. 2. The observed total neutron cross section of rhodium (Ribon et al., 1961). This measurement was made using the Saclay 25 MeV linear electron accelerator.

Fig. 3. The observed total neutron cross section of mercury (Carpenter and Bollinger, 1960). This measurement was made using the ANL fast chopper with a time-of-flight resolution of 12 nsec m⁻¹.

TABLE I

CHARACTERISTICS OF SOME POPULAR NEUTRON DETECTORS[a]

Detector	Typical thickness (cm)	Efficiency at 1 keV (%)	Practical[b] timing resln. (nsec)	Neutron peak	
				Full width at $\frac{1}{2}$ max (%)	Equivalent electron energy (MeV)
$^{10}BF_3$ counter (150 cm of Hg)	10	1	500	5	2.3
^{10}B–NaI(Tl)	2 (^{10}B) 3 (NaI)	25	$\gtrsim 3$	10	0.48
^{10}B-loaded liquid scint.	1	30	400	60	0.1
^{10}B-loaded glass scint.	1	30	not quoted	50	0.2
^{10}B–ZnS(Ag)	0.05	1	100	no peak	0.2
^{6}Li-loaded glass scint.	2.5	15–20	2	10–25	1.6
^{6}LiI(Eu) crystal	2.5	20	not quoted	12	4.1

[a] Brooks (1961).

[b] These figures represent the best performances so far achieved. From the known decay times and scintillation efficiencies of both the ^{10}B–NaI(Tl) and ^{6}Li-loaded glass, it is theoretically possible to achieve timing resolutions ~ 1 nsec.

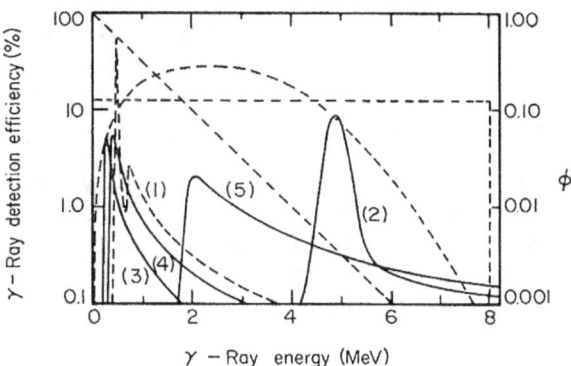

Fig. 4. Efficiency for detecting γ-rays of some neutron detectors listed in Table I (Brooks, 1961). (1) ^{10}B + NaI (Tl); (2) ^{6}LiI (Eu); (3) ^{10}B-loaded liquid; (4) ^{10}B-loaded glass; (5) ^{6}Li-loaded glass. The broken curves represent the three different assumed shapes for the γ-ray spectra.

quantitative estimates of the γ-ray sensitivity of a number of the detectors listed in Table I. Brooks assumed three different γ-ray spectra incident on the detectors: (i) a "flat" spectrum; (ii) a typical fission γ-ray spectrum; and (iii) a typical capture γ-ray spectrum from a medium or heavy nucleus. The γ-ray detection efficiency E_γ versus γ-ray energy for five different detectors is shown in Fig. 4. The ^{10}BF$_3$ counter is the most insensitive γ-ray detector and is normally used where low energy neutrons ($\gtrsim 100$ eV) are to be detected in the presence of severe γ-ray backgrounds.

The choice of detector naturally depends upon the particular requirements of an experiment. However, the trend in recent years has been away from ^{10}BF$_3$ counters and toward scintillation detectors. This is due to the need for high efficiencies in the kilovolt region coupled with the need for improved time resolutions of $\gtrsim 0.1$ μsec.

The increasing interest in nanosecond neutron time-of-flight experiments has resulted in the development of a fast NaI–^{10}B system by Good *et al.* (1958) and the ^{6}Li-loaded glass scintillator by Firk *et al.* (1961). The time resolutions reported for these two detectors are ~ 5 and ~ 2 nsec, respectively. These resolutions are obtained when observing γ-rays and must, of course, be increased appropriately to include the neutron capture time in the detector (Bollinger *et al.*, 1962).

In those applications which do not require fast time resolution, the boron-loaded liquid scintillator offers exceptionally high efficiency in the energy region of interest (Muehlhause and Thomas, 1953, and subsequent developments by the Argonne group).

III. ANALYSIS OF NEUTRON SPECTROMETER TRANSMISSION DATA

In the case of a single isotope* with nonzero nuclear spin, there are three classes of effects which must be considered in attempting an analysis:

(i) the intrinsic effects of (a) interference between resonance and potential scattering for s-wave levels and (b) the interference between levels of the same angular momentum and parity;

(ii) the mixed intrinsic-experimental Doppler effect (Section III,A,2) which can be modified to a limited extent by changing the temperature of the sample; and

(iii) the purely experimental effects of (a) lack of perfect resolution (i.e., the inability, within practical limits, of reducing the energy spread of the neutron beam to such a small extent that no significant variation of cross section occurs within the interval), and (b) lack of infinite statistical accuracy on each transmission value.

In addition, serious experimental difficulties often arise from the presence of various types of background which occur because of the detection of γ-rays and of neutrons with energies different from those under consideration. These background effects are associated with each particular experimental apparatus and cannot be discussed in general. They will therefore be omitted in our further discussion, and it will be assumed that transmission data have been derived for which the proper background corrections have been made.

The analysis of a given set of transmission data in the most general form is a formidable task and is not usually attempted. (However, with the steady increase in both the availability and the storage capacity of digital computers, this picture is slowly changing, and machine calculations are being made in which increasingly greater complications are considered simultaneously.) Situations are sought in which only a limited number of factors are present, the remainder being either negligible or included as small corrections. The most useful division is in terms of resolution. Where the resolution width is less than the level width, the shape of the observed resonance is significant and "shape analysis" (described in Section A) can be attempted. This involves a match of the experimental data, point by point, with the theoretical curves containing suitably chosen parameters. For most of the observed levels, the resolution width is larger than the level width, and the observed width is then no longer an indication of the

---

* In the general case of a mixture of isotopes, isotopic assignments of the levels must first be made before a complete analysis can be attempted; the problem then reduces to that under consideration. Nevertheless, it is frequently possible to get partial results.

level width. In general, the resolution function is not known with sufficient accuracy to disentangle its effects from the observed width. Clearly, it is difficult to deduce resonance parameters from such a resolution-broadened curve. "Area analysis," described in Section B, is then the most common form of analysis and transmission measurements on samples of several thicknesses are useful in the determination of resonance parameters.

A. Shape Analysis

1. Low Neutron Energies ($\gtrsim 1$ keV)

Except for the case of the fissionable isotopes, which is treated separately in Chapter V, the level widths below 1 keV neutron energy are generally small compared with level spacings so that level–level interference is small and can usually be neglected. This is fortunate, since the inclusion of level–level interference terms demands a knowledge of the spins J of all the levels, and such information is available only in a few favorable cases.

The only other intrinsic effect left is that of the interference between resonance and potential scattering, and this is readily included in the analysis.

Before describing some of the techniques of shape analysis, it is necessary to discuss the Doppler broadening of resonances previously mentioned in Section I,A. The following outlines the theory of this effect.

2. Doppler Broadening of the Resonance Form

In order to take into account the thermal motions of the target nuclei, the recoil of the compound nucleus must be considered. The simplest case is that of a target nucleus which is initially at rest and free to move: the resonance term given by Eq. (28) then becomes

$$\sigma_T(E_L) = \frac{\sigma_0(\Gamma/2)^2}{[(E_L - R) - E_\lambda]^2 + (\Gamma/2)^2} \tag{35}$$

where E_L is the laboratory energy of the incident neutron, and R is the recoil energy of the compound nucleus such that

$$R = mE_L/(m + M) \tag{36}$$

where m is the mass of the neutron, and M the mass of the target nucleus. Note that Eq. (35) is usually written with $R = 0$ (i.e., the system in which the compound nucleus is at rest).

In practice, the target nuclei are not at rest but have a distribution of velocities characteristic of their environment and its temperature. Bethe and Placzek (1937) modified Eq. (35) for the case in which the target nuclei are treated as a classical gas. It is necessary to include in the energy dependence of Eq. (35) the relative motion of the target nucleus and incident neutron and then to average over the Maxwellian distribution of velocities of the atoms. Their result is obtained as follows.

In general, the "Doppler-broadened" cross section $\sigma_\Delta(E_L)$ may be written in terms of an energy transfer function $S(E_t) = S(E_L - E)$ which is convoluted with $\sigma(E)$ to give*

$$\sigma_\Delta(E_L) = \int S(E_L - E) \cdot \sigma(E) \, d(E_L - E) \qquad (37)$$

The form of the function $S(E_L - E)$ depends on the model which is assumed for the medium in which the target nuclei are bound. For a classical gas, the energy transfer $(E_L - E)$ is

$$(E_L - E) \approx R + mvw_x \qquad (38)$$

where v is the laboratory energy of the neutron, and w_x is the velocity of the target nucleus in the direction of the incident neutron beam. The distribution of velocities w_x is given by the Maxwell–Boltzmann formula

$$p(w_x) \, dw_x = M/(2\pi MkT)^{1/2} \exp\{-Mw_x^2/2kT\} \, dw_x \qquad (39)$$

where T is the gas temperature, and k is the Boltzmann constant. The function $S(E_L - E)$ then becomes

$$S(E_L - E) = (1/\Delta \sqrt{\pi}) \exp\{-(E_L - E - R)^2/\Delta^2\} \qquad (40)$$

where

$$\Delta = 2(RkT)^{1/2}$$

is the "Doppler width." Jackson and Lynn (1962) point out that the same result is obtained for a classical solid in which the nuclei are considered as linear harmonic oscillators whose energies are described by Boltzmann statistics.

Lamb (1939) calculated the shape of a resonance for target nuclei bound in a quantum mechanical crystal. He discussed the oscillation of a target nucleus bound in a harmonic oscillator well (HOW). The energy of the oscillation is then one of the eigenvalues of the HOW, and the expectation value of this energy is equal to the recoil energy R: There is a finite probability of the oscillator remaining in its initial state [cf. the Mössbauer effect (1958)].

---

* Note that the Doppler broadening acts on the cross section, whereas the broadening arising from lack of perfect resolution acts upon the transmission.

For nuclei bound in a Debye crystal, Lamb calculated the modes of oscillation and the resulting resonance shape using dispersion theory.

The conclusions were:

(i) for a cold crystal (relative to the Debye temperature θ_D) and for a narrow resonance width and low recoil energy (both relative to $k\theta_D$), the "recoilless" peak is expected. Some structure may also appear at higher energies due to the creation or annihilation of one or more phonons in the crystal lattice.

(ii) at high crystal temperature (i.e., weak binding such that $\Delta + \Gamma \gg 2k\theta_D$), the result is the same as that of the classical gas broadening outlined above except that the mean energy per degree of freedom $(\frac{1}{2}kT)$ must be replaced by the quantum-mechanical mean energy

$$\bar{\epsilon} = \frac{1}{2} \int_0^{\nu_m} \coth(h\nu/2kT)g(\nu)h\nu \, d\nu \tag{41}$$

where $g(\nu)$ is the frequency spectrum of the lattice, and ν_m is the maximum frequency in the spectrum. (Generally, $\bar{\epsilon} > \frac{1}{2}kT$ but approaches it at high temperature.)

Landon (1954) verified the classical gas approximation by studying the resonance profile of the 1.26 eV resonance in rhodium as a function of temperature. Jackson and Lynn (1962) have carried out measurements on ^{189}Os, ^{238}U, and U_3O_8 for a range of sample temperatures down to 4°K. For the metal samples, they obtained good agreement with their data using Lamb's theory and a simple Einstein frequency spectrum for the lattice.

The Doppler broadened cross section corresponding to Eq. (28) is usually written

$$\sigma_T(\beta,x) = \sigma_0\Psi(\beta,x) + \sigma_0 \tan(2a/\lambda)\Phi(\beta,x) + \sigma_{\text{pot}} \tag{42}$$

where

$$\Psi(\beta,x) = (\beta\sqrt{\pi})^{-1} \int_{-\infty}^{\infty} (1 + y^2)^{-1} \exp\{-(x - y)^2/\beta^2\} \, dy \tag{43}$$

and

$$\Phi(\beta,x) = (\beta\sqrt{\pi})^{-1} \int_{-\infty}^{\infty} [y/(1 + y^2)] \exp\{-(x - y)^2/\beta^2\} \, dy \tag{44}$$

Here,

$$\beta = 2\Delta/\Gamma, \qquad x = (2/\Gamma)(E - E_\lambda) \qquad \text{and} \qquad y = (2/\Gamma)(E_{\text{CM}} - E_\lambda)$$

(E_{CM} is the center-of-mass energy of the neutron).

Other dimensionless parameters used in the literature in place of β are

$$\xi \equiv \Gamma/\Delta = 2/\beta: \qquad t = (\Delta/\Gamma)^2 = \tfrac{1}{4}\beta^2$$

Extensive tables of the function $\psi(\beta,x)$ have been published (Rose *et al.*, 1953). However, since most analysis is currently performed using computers, these functions are computed as needed.

3. Some Techniques of Shape Analysis

The "wings" of a resonance are usually unaffected by resolution and Doppler broadening since the cross section is only varying slowly with energy in this region. The Breit–Wigner formula Eq. (28) may therefore be applied directly. A least-squares analysis frequently gives useful combinations of resonance parameters. Specifically, the cross section on the wings of a resonance is

$$\sigma_{n,T,\text{wings}} = \frac{\sigma_0\Gamma^2}{4(E - E_\lambda)^2} + \frac{\sigma_0\Gamma(a/\lambda)}{E - E_\lambda} + \sigma_{\text{pot}} \qquad (45)$$

Assuming that the resonance energy E_λ can be obtained adequately by inspection, a curve fit to the data gives σ_{pot}, $\sigma_0\Gamma^2$ (proportional to $g\Gamma_n\Gamma$), and $\sigma_0\Gamma a$ (where σ_0 is proportional to $g\Gamma_n$). The main requirement here is that the wings do not include significant contributions from other levels.

The central region of a resonance, on the other hand, is usually affected by both types of broadening. All the methods of shape analysis used for the central region have been a variation of the "trial and error" method. Here, a preliminary analysis or inspection is used to give approximate values of the parameters. The Doppler-broadened cross section, transmission and resolution-broadened transmission are then computed successively and are compared with the experimental transmission values. An inspection shows the changes that are necessary to the input values. A new set of corrected transmission values are then calculated, and comparison with observations again made. The whole process is repeated until a "satisfactory" fit is obtained. Early applications of this method were made by Havens and Rainwater (1946), McDaniel (1946), and Meyer (1949).

Sailor (1953) introduced an improved method for finding the corrections to be applied to the initial choices of the parameters. The transmission function is expanded in a Taylor series about the values determined by the initial parameters, and terms are kept only to the first order in the corrections for σ_0 and Γ. The method of least-squares is then used to find those values of the corrections which reduce to a minimum, the sum of the squares of the differences between observed and calculated transmission values. This process can be repeated until it is shown that the higher terms in the Taylor expansion are indeed negligible. Seidl *et al.* (1954) have developed a similar method of analysis.

Until recently, the above methods could be applied in a few special cases in which only one low-lying level was adequately resolved in a given isotope. With the recent increase in spectrometer resolution, there are now frequent cases in which many levels in a given isotope are sufficiently well resolved for a shape analysis to be applied. Desk calculation methods become inadequate and digital computers must be used. Harvey and Atta (1961) have set up schemes for the IBM 704 and IBM 7090 computers which can shape-analyze a transmission curve for as many as six levels at once. The method is similar in principle to the Sailor procedure in that transmission values for assumed input parameters are calculated, and least-squares solutions made to calculate corrections to the input parameters. However, the program is of far larger scope: up to six resonances may be analyzed together and iteration is made to a much higher degree of precision. Certain other refinements, to be discussed below, are also included.

In the Harvey–Atta method, the cross section is represented as a sum of single-level Breit–Wigner resonances including interference between resonance and potential scattering, but not including interference between resonances. Doppler and resolution broadening are both represented by Gaussian functions. The resolution width can be of the form $R_0 + R_1 i$ (i = channel number) to take into account an energy-dependent part of the resolution such as effects of burst and channel width, and an energy dependent part, which might arise from a finite detector depth or variation in neutron moderation time in the source. Provision is made for a superimposed transmission variation of the form

$$P(E) = K_0 + (K_1/\sqrt{E}) + (K_2/E) \tag{46}$$

which may arise from contributions of resonance levels outside the energy region being analyzed. This variation can also take into account a variation of the neutron flux with energy when the program is used to analyze partial cross section data (such as capture or scattering). The least-squares fit then gives values of the resonance energy, the full width, and the product $fg\Gamma_n^0$ where f is the fractional abundance of the isotope to which the resonance belongs; g is the statistical weight factor; and Γ_n^0 is the reduced neutron width. The program includes estimates of the errors in the finally-selected values of the parameters.

An accompanying program performs area analyses (Section III,B) on as many as thirty resonances at once. The parameters deduced from the area analysis may then be used as initial estimates for the shape analysis program.

An additional advantage of the Harvey–Atta procedure is that a large amount of overlap of levels can be tolerated. This is due to the summation of the contributions of all the levels in the energy interval under consideration. Such a procedure is valid provided that level–level interference is negligible.

Several other specialized approaches to the shape analysis problem are discussed by Melkonian (1956) and Michaudon (1961).

4. ENERGY $\gtrsim 1$ keV, NO DOPPLER BROADENING AND GOOD RESOLUTION

The improved resolution of neutron spectrometers during recent years has resulted in an increasing number of total cross section measurements above 1 keV. In these cases, levels with large neutron scattering widths compared with the level spacings are frequently encountered so that interference effects between levels of the same spins and parities become important. It is then necessary to use the general multilevel form for the cross section as given in Eqs. (25) and (26).

Measurements of this type were first reported by Bollinger et al. (1955) who determined the total cross section of manganese from thermal energies to ~10 keV. A theoretical analysis of these measurements was made by Krotkov (1955).

More recently, the total cross section of vanadium has been measured up to 25 keV by Firk et al. (1963b) and the data analyzed using the Thomas approximation given in Eq. (26). The observed total cross section is shown in Fig. 5. The energies, total widths and spins [see Eq. (80)] of each level are determined from a shape analysis of the observed total cross section. The shape analysis is possible since the effects of resolution and Doppler broadening are negligible at almost all points of the measurement. The parameters giving the best fit to the data (obtained from a least-squares analysis programmed for an IBM 704 computer) are shown in Table II.

TABLE II

BEST VALUES OF RESONANCE PARAMETERS FOR THE
NEUTRON CROSS SECTION OF $^{51}V^a$

E (keV)	J	$\gamma_{\lambda n}^2$ (keV)	Γ_λ (keV)
4.169 ± 0.005	4	3.304 ± 0.035	0.508 ± 0.006
6.886 ± 0.010	3	6.475 ± 0.070	1.28 ± 0.014
11.810 ± 0.004	3	21.25 ± 0.18	5.50 ± 0.05
16.6 ± 0.1	4	1.13 ± 0.03	0.35 ± 0.10
17.4 ± 0.1	4	0.3 ± 0.1	0.09 ± 0.03
22.32 ± 0.03	3	2.68 ± 0.11	0.95 ± 0.04

^{a} Firk et al. (1963b).

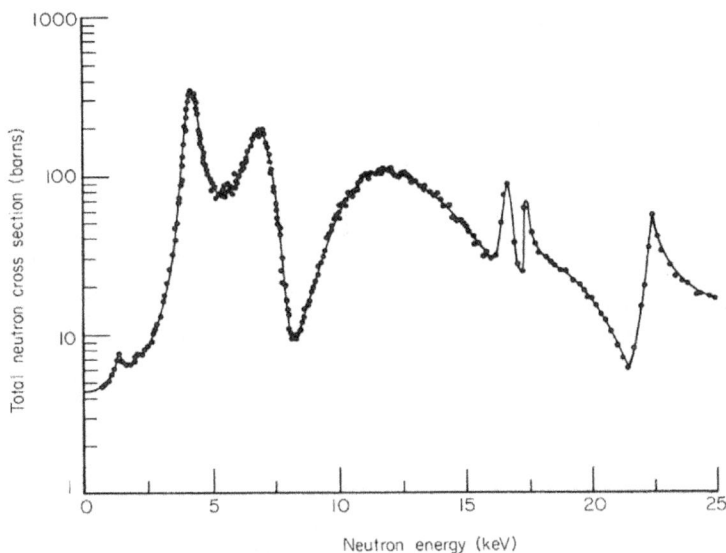

Fig. 5. The observed total neutron cross section of vanadium up to an energy of 25 keV (Firk *et al.*, 1963b).

A feature of the analysis of the vanadium total cross section is the inclusion of the effect arising from distant levels. The following method is used to determine this effect. In the low energy approximation of Eq. (32) the thermal scattering length a_J for spin J may be written

$$a_J = a(1 - R_{nnJ}^{\infty}) \tag{47}$$

$$= a \left(1 - \sum_\lambda (\gamma_{\lambda nJ})^2/E_{\lambda J} - R_{nnJ}^{\infty'} \right) \tag{48}$$

where a is the nuclear radius, $\Sigma_\lambda(\gamma_{\lambda nJ})^2/E_{\lambda J}$ is the contribution at thermal energy ($E \approx 0$) due to levels of spin J in the energy range 0–25 keV and $R_{nnJ}^{\infty'}$ is the cumulative effect of levels of spin J outside the range 0–25 keV. From the thermal scattering data of Peterson and Levy (1952), two pairs of values of a_J can be obtained, and therefore two pairs of values of $R_{nnJ}^{\infty'}$. The best fit to the data is obtained for both $R_{nnJ=3}^{\infty'}$ and $R_{nnJ=4}^{\infty'}$ negative. The sensitivity of the fit to the correct choice of $R_{nnJ}^{\infty'}$ is clearly shown in Fig. 6 in which $R_{nnJ}^{\infty'}$ is considered a constant. In a final analysis of the data, $R_{nnJ}^{\infty'}$ is resolved into an energy dependent part (due to levels outside the range 0–25 keV which are, nevertheless, close enough to exert a mild energy dependence) plus a constant. The theoretical implications of the

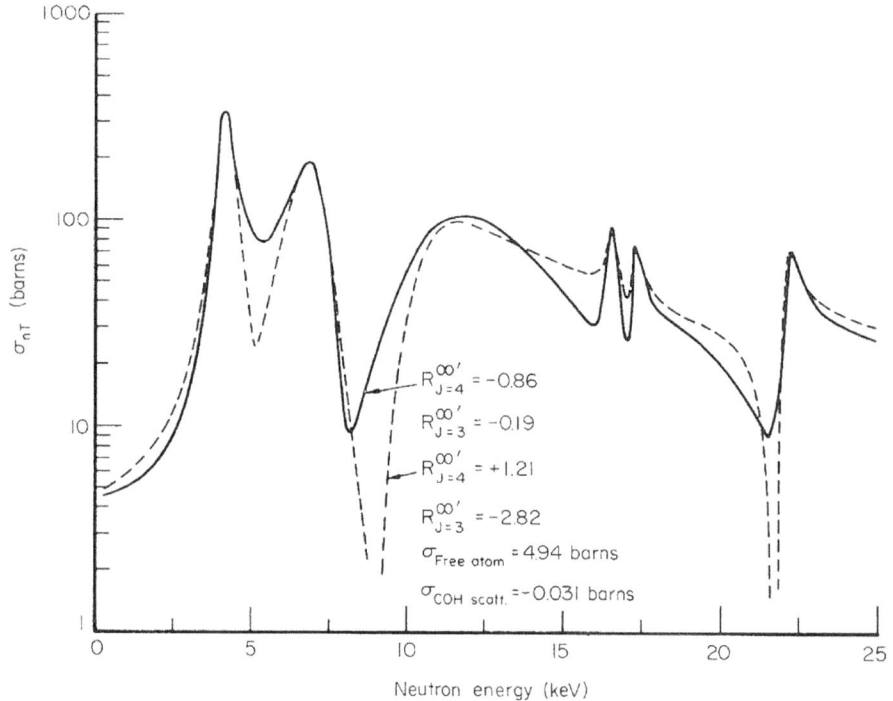

INITIAL PARAMETERS

E_λ (eV)	Γ_λ (eV)	J	E_λ (eV)	Γ_λ (eV)	J
4,150	540	4	16,600	190	4
6,800	1200	3	17,400	200	4
11,700	4400	3	22,300	950	3

FIG. 6. A multilevel fit to the vanadium total cross section using constant values of $R_{nnJ}^{\infty\prime}$. The pairs of values of $R_{nnJ}^{\infty\prime}$ are consistent with the thermal scattering cross section data. The resonance energies, total widths, and spins of the six levels were obtained initially by inspection. Final values of the resonance parameters deduced from a least-squares analysis of the cross section are given in Table II (Firk *et al.*, 1963b).

vanadium results are discussed in the paper of Firk *et al.* (1963b) and in Section V.

B. Area Analysis

The most widely used method of analyzing neutron transmission data to determine resonance parameters is the "area-method" developed by

Havens and Rainwater (1946, 1951). The usefulness of this method arises from the fact that the total area under an isolated transmission dip (plotted as a function of energy, say), is independent of the resolution function of the neutron spectrometer. A quantitative treatment of the validity of the area method of analysis has been given by Lynn and Rae (1957).

The expression for the area under a transmission dip in terms of the resonance parameters and sample thickness may be obtained using the single-level Breit–Wigner form for the total cross section given by Eq. (28). This form is suitable in most cases of analysis met in the energy range below 50 keV. Multilevel interference effects have been considered by Lynn (1958) and these may be included where necessary.

Using Eq. (28) the transmission is

$$T(x) = \exp\{-n\sigma_T(x)\} = \exp\left\{ -\frac{n\sigma_0}{1 + x^2} - \frac{n\sigma_0 x}{1 + x^2} \tan(2a/\lambda) - 4\pi n a^2 \right\}$$

(49)

The slowly varying contribution to the transmission from nearby and distant levels may be included in the term $\exp(-4\pi n a^2)$ where a now becomes an effective scattering length [cf. R' of Eq. (33)].

Writing $T_P = \exp(-4\pi n a^2)$ the area under a transmission dip (in energy units) is defined as

$$A_E(\infty) = \int_{-\infty}^{\infty} [1 - T(E)/T_P]\, dE$$

(50)

or, in the x-notation

$$A_x(\infty) = \int_{-\infty}^{\infty} [1 - T(x)/T_P]\, dx = (2/\Gamma)A_E$$

(51)

1. No Doppler Broadening and No Interference Term

If the interference term in Eq. (28) is neglected (valid for $n\sigma_0 \gtrsim 50$) and Doppler broadening is negligible, then the area function A_x' (say) is given by

$$A_x'(\infty) = \int_{-\infty}^{\infty} \left[1 - \exp\left(\frac{-n\sigma_0}{1 + x^2} \right) \right] dx$$

(52)

Substituting $x = \tan(\phi/2)$ this becomes

$$A_x'(\infty) = (n\sigma_0/2) \exp(-n\sigma_0/2) \left\{ \int_{-\pi}^{\pi} \exp(-n\sigma_0 \cos \phi/2)[1 - \cos \phi]\, d\phi \right\}$$

(53)

which is readily shown to be

$$A_x'(\infty) = \pi n\sigma_0 \exp(-n\sigma_0/2)[I_0(n\sigma_0/2) + I_1(n\sigma_0/2)]$$

(54)

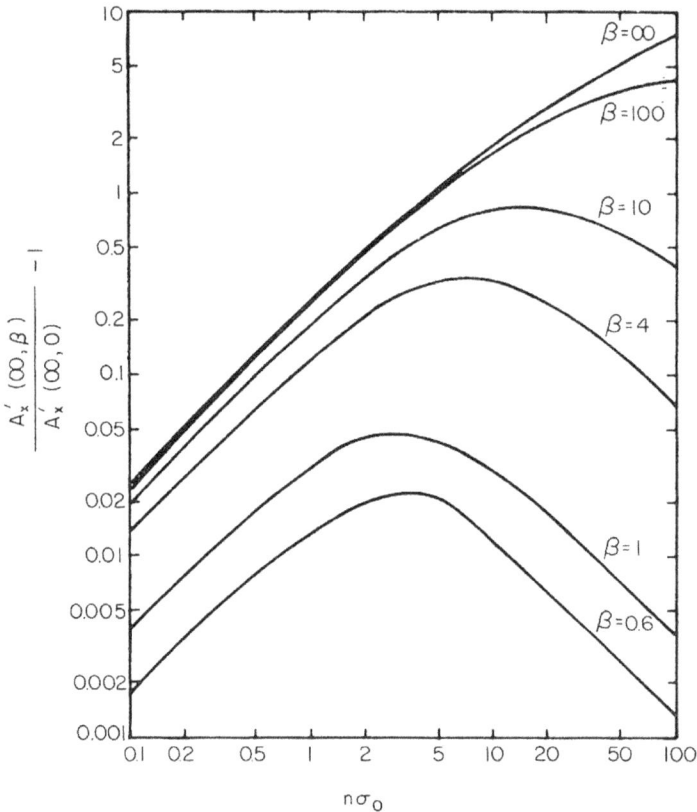

FIG. 7. A curve of the relative Doppler broadening of the area function $A_x'(\infty,\beta)$ as a function of $n\sigma_0$ (Melkonian et al., 1953).

where I_0 and I_1 are Bessel functions of imaginary argument of order 0 and 1, respectively. It is interesting to note that the analytical expression for $A_x'(\infty)$ was first given by Ladenburg and Reiche (1913), in connection with the absorption by the earth's atmosphere, of spectral lines from the sun.

The asymptotic values for Eq. (54) are, in the energy notation

$$\lim_{n\sigma_0 \to 0} A_E'(\infty) = \pi n\sigma_0 \Gamma/2 \tag{55}$$

and

$$\lim_{n\sigma_0 \to \infty} A_E'(\infty) = (\pi n\sigma_0)^{1/2}\Gamma \tag{56}$$

Equations (55) and (56) represent the "thin" and "thick" sample approximations, respectively, and have been widely used to determine σ_0 and Γ by careful choice of sample thickness. Melkonian et al. (1953) have discussed

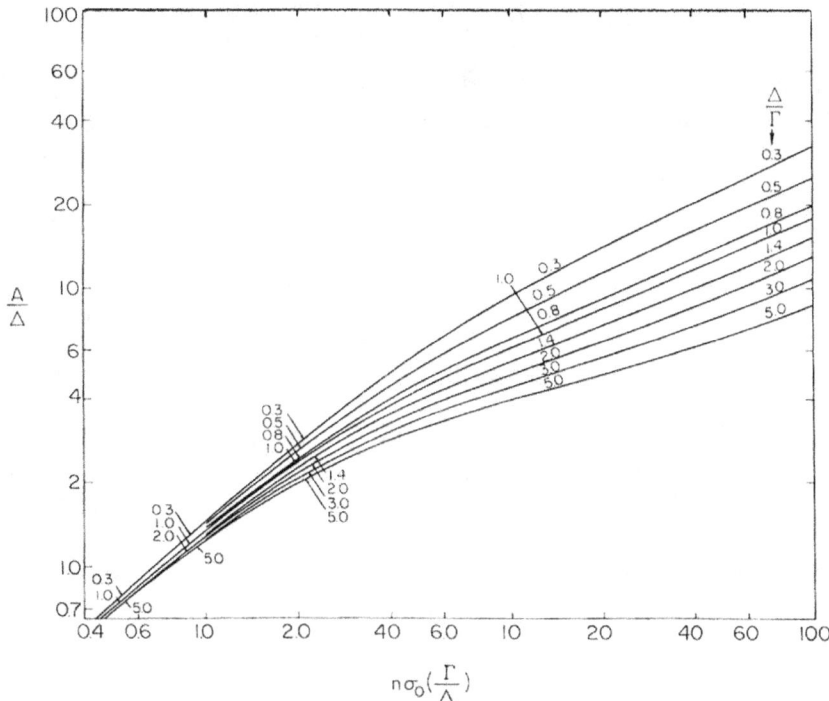

FIG. 8. A set of curves of $A_E'(\infty,\beta)/\Delta$ as a function of $n\sigma_0(\Gamma/\Delta)$ for a range of values of Δ/Γ. These curves are used in the Hughes' method of area analysis (Hughes, 1955a).

the problem for sample thicknesses intermediate in value between the limits imposed in Eqs. (55) and (56). These authors show that for sample thicknesses such that $1 \gtrsim n\sigma_0 \gtrsim 10$ the area is given by

$$A_E'(\infty) \propto \sigma_0 \Gamma^P \qquad \text{where} \quad 2 > P > 1$$

2. Doppler Effect but No Interference Term

The effect of Doppler broadening on the total cross section in the neighborhood of a resonance and the consequent change in the area under a transmission dip was first considered by Melkonian et al. (1953). In their treatment, the interference term was neglected.

On averaging over the thermal velocity distribution of the atoms in the sample the resonance term $\propto (1 + x^2)^{-1}$ becomes

$$(1 + x^2)^{-1} \rightarrow \Psi(\beta,x) = (\beta\sqrt{\pi})^{-1} \int_{-\infty}^{\infty} (1 + y^2)^{-1} \exp\{-(x - y)^2/\beta^2\}\, dy$$

where the terms are defined in Section III,A,2.

The Doppler broadened area function is now written

$$A_x'(\infty,\beta) = \int_{-\infty}^{\infty} \{1 - \exp[-n\sigma_0\Psi(\beta,x)]\}\, dx \qquad (57)$$

The integral in Eq. (57) has been numerically integrated (Melkonian *et al.*, 1953) and a curve of the relative Doppler broadening of the area as a function of $n\sigma_0$ presented. Their curve is shown in Fig. 7. It is seen that the effect of Doppler broadening is to increase the range of validity of the thick sample approximation. In this treatment, it is also possible to correct simultaneously for the "wing areas" excluded by the practical need to limit the measured area to a finite interval about the resonance energy.

A useful method of area analysis due to Hughes (1955a) enables corrections for Doppler broadening and finite sample thickness to be made simultaneously. Corrections for the excluded wing areas must, however, be made by successive approximations. Figure 8 gives a set of curves of $A_E'(\infty,\beta)/\Delta$ as a function of $n\sigma_0(\Gamma/\Delta)$ for a wide range of (Δ/Γ). A typical example readily demonstrates the Hughes' method. The analysis of the data given by Firk (1958) for the 23.9 eV resonance in tantalum proceeds as follows.

The Doppler width $\Delta = 0.116$ eV

Sample thicknesses (atoms/b)	Observed area $A_E'(\infty,\beta)$ (eV)
0.000304	0.142 ± 0.010
0.00076	0.279 ± 0.011
0.00954	0.915 ± 0.027

For each sample thickness $A_E'(\infty,\beta)/\Delta$ is a constant. Using Hughes' curve shown in Fig. 8, a table is constructed for sample thickness $n = 0.000304$ atom/b (here, $A_E'(\infty,\beta)/\Delta = 1.22$). The details are given in Table III.

This procedure is repeated for the different sample thicknesses, and the

TABLE III

DETERMINATION OF THE FUNCTIONAL RELATIONSHIP $\sigma_0 = f(\Gamma)$ FOR THE 23.9 eV RESONANCE IN Ta USING HUGHES' METHOD[a]

Δ/Γ	Γ (eV)	$n\sigma_0(\Gamma/\Delta)$	$n\sigma_0$	σ_0 (b)
0.2	0.58	0.81	0.162	533
0.4	0.28	0.84	0.336	1105
0.7	0.166	0.87	0.609	2003
1.0	0.116	0.90	0.900	2960
2.0	0.058	0.94	1.88	6184

[a] Firk (1958).

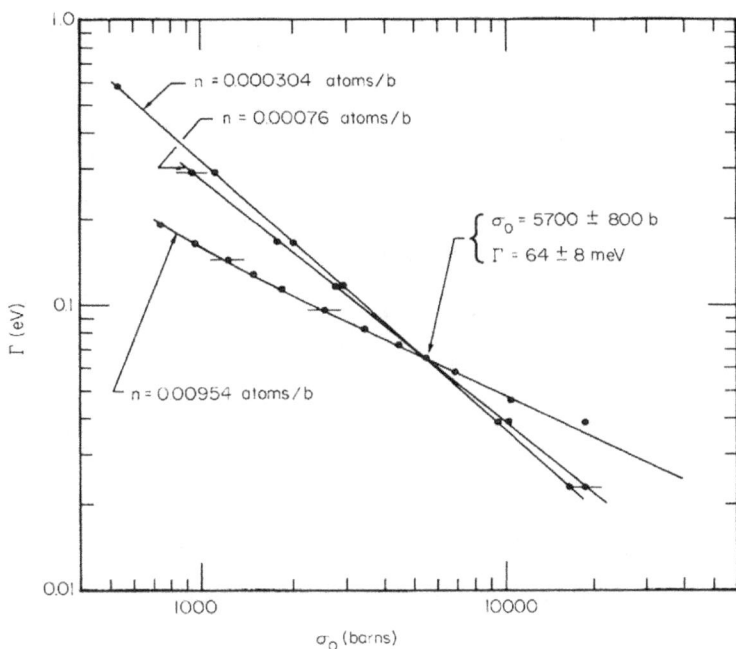

FIG. 9. The results of an area analysis of the 23.9 eV resonance in Ta (Firk, 1958) Three different sample thicknesses were used in order to determine σ_0 and Γ.

values of σ_0 and Γ are then determined from a plot of $\log \sigma_0$ versus $\log \Gamma$ as shown in Fig. 9. The results are

$$\sigma_0 = 5700 \pm 800 \quad \text{b}; \qquad \Gamma = 64 \pm 8 \quad \text{meV}$$

3. INCLUSION OF THE INTERFERENCE TERM

a. No Doppler Effect. The asymmetry due to interference between resonance and potential scattering is clearly demonstrated in the transmission curves shown in Fig. 10 (Lynn *et al.*, 1958b). As the sample thickness increases the contribution to the area above the line, T_P becomes more and more pronounced. The sign of the area [as defined in Eq. (50)] above the line $T/T_P = 1$ (positive) is opposite to that of the area below $T/T_P = 1$. The result is that the area is decreased by the interference effect.

The method of area analysis including the interference effect was pioneered by Lynn (1958) whose method is outlined below.

The area function including the interference term may be written

$$A_x(\infty,\beta) = A_x{}'(\infty,\beta) - \sum_{m=1}^{\infty} C_m(\beta) \qquad (58)$$

34

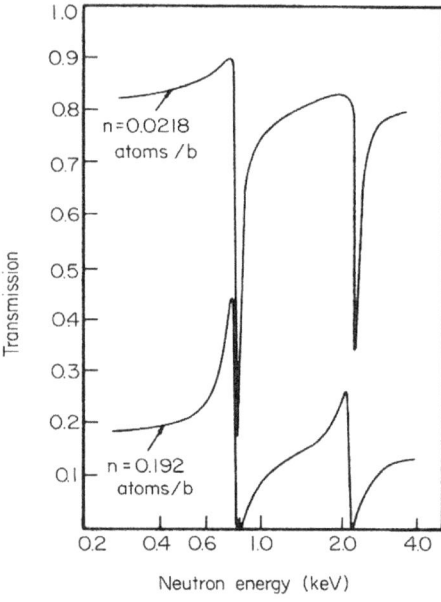

FIG. 10. An example of interference between resonance and potential scattering for different values of sample thickness. The resonances are those observed in the cross section of bismuth (Lynn et al., 1958).

where the positive quantity $C(\beta) = \Sigma_m C_m(\beta)$ takes into account the net decrease in the total area. In the case of no Doppler broadening ($\beta = 0$), Lynn gives the following expression for $C_m(0)$:

$$C_m(0) = \pi n\sigma_0 \exp\left(\frac{-n\sigma_0}{2}\right) \frac{(2m-3)!!}{(2m)!} \left(\frac{2n\sigma_0 a^2}{\lambda^2}\right)^m \left[I_{m-1}\left(\frac{n\sigma_0}{2}\right) + I_m\left(\frac{n\sigma_0}{2}\right) \right] \quad (59)$$

For energies above ~ 10 keV* terms involving a/λ explicitly should be replaced by $\frac{1}{2}\tan(2a/\lambda)$.

Lynn (1960a) has made extensive calculations of a natural area function defined by

$$A_N(\infty,0) = \tfrac{1}{4}\tan(2a/\lambda)A_x(\infty,0) \quad (60)$$

These calculations are available (Lynn, 1960b) for $n\sigma_0$ in the range 1–1024 and $\frac{1}{2}n\sigma_0 \tan^2(2a/\lambda)$ in the range 0.02–2.7.

---

* Seth (1959), using analytical methods similar to Lynn, has discussed a method of area analysis, including the interference term but excluding Doppler broadening, for neutron energies in the range 10–1000 keV ($a/\lambda > 0.1$). This energy range is, however, mostly outside the scope of the present article.

b. Doppler Effect Included. The expression for the area function $A_x(\infty,\beta)$ including Doppler broadening of both the resonance and interference terms is

$$A_x(\infty,\beta) = \int_{-\infty}^{\infty} \{1 - \exp[-n\sigma_0\Psi(\beta,x)]\} \, dx$$

$$- \sum_{m=1}^{\infty} \frac{(n\sigma_0 \tan(2a/\lambda))^{2m}}{(2m)!} \int_{-\infty}^{\infty} \Phi^{2m}(\beta,x)$$

$$\times \exp[-n\sigma_0\Psi(\beta,x)] \, dx \qquad (61)$$

where

$$\Phi(\beta,x) = (\beta\sqrt{\pi})^{-1} \int_{-\infty}^{\infty} y/(1 + y^2) \exp\{-(x - y)^2/\beta^2\} \, dy$$

Lynn (1960b) has calculated $A_x(\infty,\beta)$ by numerical integration methods and has presented tables of the area function:

$$A_N(\infty,\beta) = \tfrac{1}{4} \tan(2a/\lambda) A_x(\infty,\beta) \qquad (62)$$

for wide ranges of $\beta(1(\sqrt{2})64)$ and $n\sigma_0 \tan^2(2a/\lambda)$ $(0.014865(\sqrt{2})6.4)$.

c. Determination of Resonance Parameters Using Lynn's General Area Function. Before comparing an experimentally obtained area with a theoretical area, in order to determine the resonance parameters, the following corrections are necessary:

(i) A wing correction to the area $(\Delta A_E(\epsilon)$, say), due to finite energy limits.
(ii) A wing correction to the area $(\delta A_E(\epsilon)$, say), due to asymmetry of the transmission curve.
(iii) An evaluation of the effective potential transmission T_p.
(iv) An evaluation of the true resonance energy E_λ.

These corrections have been discussed in detail (Lynn *et al.*, 1958b; Lynn, 1960a). An important feature emerges, namely, that the above four correction terms can be expressed as functions of the parameters $n\sigma_0 \tan^2(2a/\lambda)$ and $\Gamma/\tan(2a/\lambda)$. The theoretical area function is also dependent upon these two quantities. The two parameters $n\sigma_0 \tan^2(2a/\lambda)$ and $\Gamma/\tan(2a/\lambda)$ are, therefore, regarded as the resonance parameters to be determined from the experimental data. (Note that there are three unknowns: σ_0, Γ, and a.) The method of analysis is therefore to assume a set of values of $n\sigma_0 \tan^2(2a/\lambda)$ and $\Gamma/\tan(2a/\lambda)$ which cover a large area of the $n\sigma_0 \tan^2(2a/\lambda)$, $\Gamma/\tan(2a/\lambda)$ plane in the region where the true values of these parameters are expected to lie. The above four correction terms are cal-

36

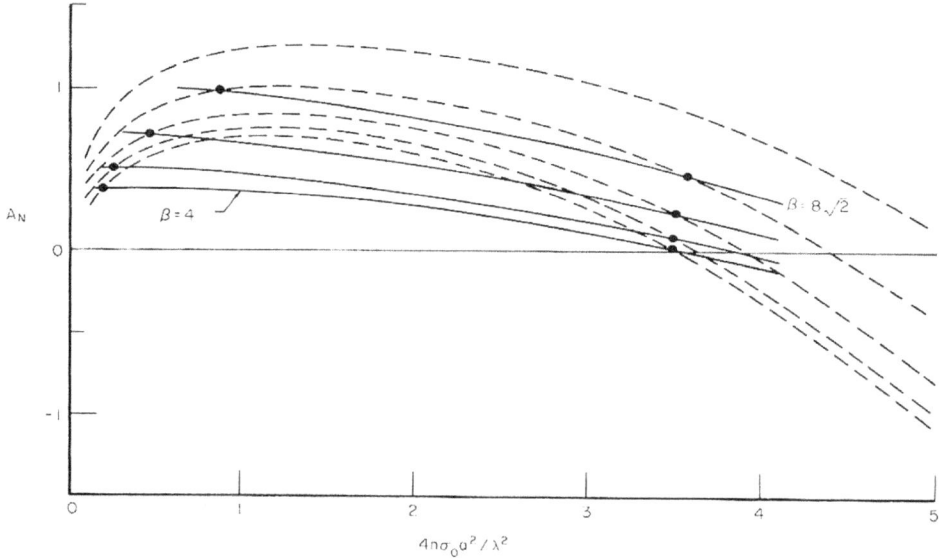

Fig. 11. An example of Lynn's general area method of analysis. A comparison of experimental and theoretical area functions for the 348 eV resonance in the cross section of ^{238}U (Firk et al., 1963a). Note the negative value of the area for large values of $n\sigma_0$. (- - -, theoretical; —, experimental data; $n = 0.0605$ atoms/b.)

culated for each pair of values in the set. Using these correction terms the following pseudocorrected area is calculated from the experimental data:

$$A_E^{\text{exp}} = \left[\sum_i \Delta E_i - \sum_i T_i \Delta E_i / T_p \right]_{E_1}^{E_2} + \Delta A_E(\epsilon) - \delta A_E(\epsilon) \qquad (63)$$

where ΔE_i is the width of an energy channel, and T_i the observed transmission associated with a point i located between the energies E_1 and E_2.

The pseudocorrected experimental areas A_E^{exp} are now compared with the theoretical areas obtained using the same set of assumed parameters. If the correct resonance parameters have been chosen for the wing corrections, potential transmission, and true resonance energy, then A_E^{exp} should agree with the theoretical area function A_E, computed from these parameters.

Agreement is obtained not only for the true parameters but also for other values of the resonance parameters which correspond to a functional relation

$$\tfrac{1}{4}\sigma_0 \tan^2 (2a/\lambda) = f(2\Gamma/ \tan (2a/\lambda)) \qquad (64)$$

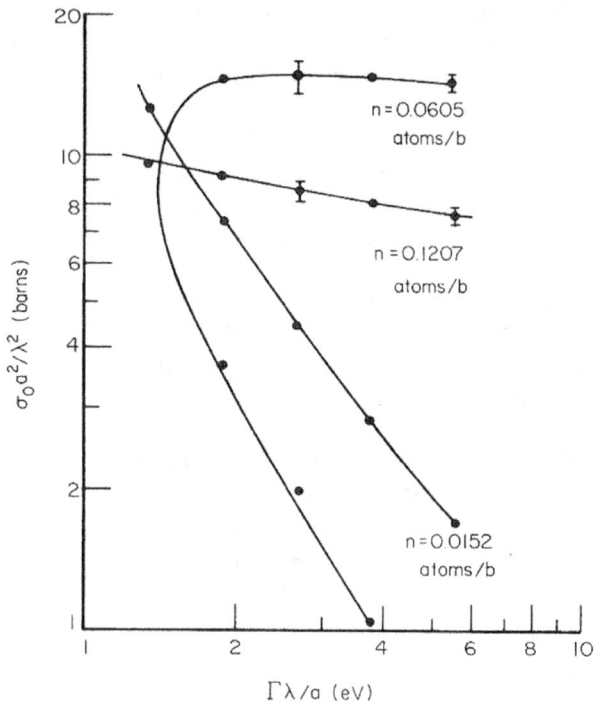

FIG. 12. A determination of the resonance parameters (σ_0, Γ, and a) for the 348 eV resonance in ^{238}U $+ n$ using three "thick" samples ($n\sigma_0 \gg 10$) (Firk et al., 1963a).

This gives rise to a self-consistent curve for a given resonance and sample thickness.

By using transmission data obtained from a range of sample thicknesses a set of self-consistent curves may be obtained. These curves will intersect at the correct values of the resonance parameters within the experimental errors.

As an example of the above method of analysis, consider the data obtained for the 348 eV resonance in ^{238}U (Firk et al., 1963a). A comparison between the pseudocorrected experimental data and the corresponding theoretical data is shown in Fig. 11. Samples containing $n = 0.01524$, 0.06045, and 0.12072×10^{24} atoms/cm² were used. The self-consistent curves from Fig. 11 are then plotted in Fig. 12, and the resonance parameters ($\sigma_0 a^2$ and Γ/a) determined from the intersection of the three curves (using a least-squares analysis).

d. Determination of Resonance Parameters Using the Harvey–Atta Method of Area Analysis. Recently, there has been a considerable increase in the number of resonances resolved below an energy of about 1 keV. The analysis of such data is no longer practicable using hand calculations which treat each resonance separately. In addition to the shape analysis method, Harvey and Atta have produced a computer program which will analyze up to thirty resonances simultaneously using an area method (Harvey and Atta, 1961). Their method has not been developed in such a general way as that due to Lynn: for instance, there are no trigonometrical factors included in the expressions for the transmission and the inclusion of level–level interference would require considerable changes in the program. Nevertheless, the method is valuable for analyzing transmission data for those nuclides in which the average level spacing is large compared with the average width.

The computer calculates the following Doppler broadened cross section and resolution broadened transmission

$$\sigma_\Delta(E') = (\Delta\sqrt{\pi})^{-1} \int_0^\infty \sigma(E'') \exp - \{(E' - E'')/\Delta\}^2 \, dE'' \tag{65}$$

and

$$T(E_i) = (R\sqrt{\pi})^{-1} \int_0^\infty \exp - \{n\sigma_\Delta(E')\} \exp - \{(E_i - E')/R\}^2 \, dE' \tag{66}$$

where all the terms in Eqs. (65) and (66), except R, have been defined previously. Here, R is the full width at half maximum of the resolution function which is assumed to be a Gaussian.

Estimates of the resonance parameters are fed into the analysis program and the theoretical areas under all the transmission dips are then calculated. Iterations are made until the computed experimental areas and theoretical areas agree within a specified accuracy. The "optimum" values of the resonance parameters are thus determined. As shown in Section III,B,2, the use of samples of different thicknesses yields values of $g\Gamma_n$ and Γ for each resonance.

Two important features of the method are:

(i) The ability to analyze many resonances simultaneously by including the cumulative effects of neighboring resonances on each other.

(ii) The inclusion of resolution-broadening in the expression for the transmission. This procedure is necessary in cases in which the wings of resonances are still affected by resolution and neighboring resonances are closely spaced. An example of the simultaneous area

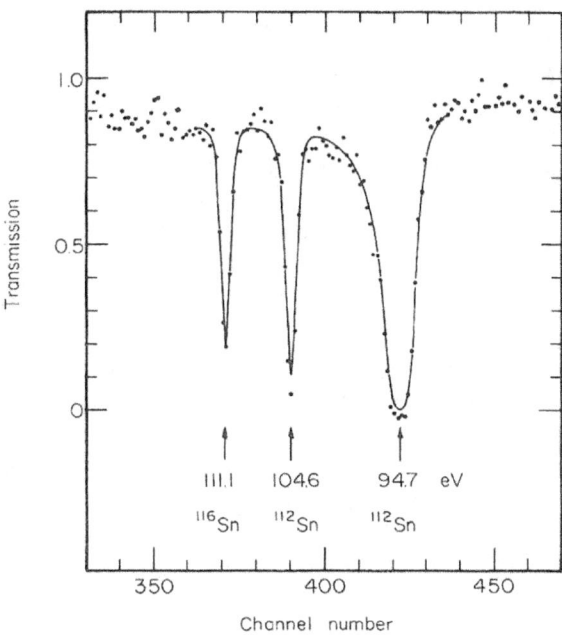

Fig. 13. An example of the simultaneous analysis of three resonances in tin using the Harvey–Atta area analysis method (Harvey and Atta, 1961; Khan and Harvey, 1962).

analysis of resonances in tin is shown in Fig. 13 (Khan and Harvey, 1962).

C. Average Cross Sections

Although the resolution of neutron spectrometers has been greatly improved during recent years, it is still not possible to resolve individual resonances in heavy elements for energies greater than a few kilovolts. Under these circumstances the average total cross section or transmission may be used to obtain the average properties of the resonances, i.e., the s- and p-wave strength functions.

Early measurements of s-wave strength functions using the average transmission technique (Hughes and Zimmerman, 1956; and Gayther and Nicholson, 1957) were limited in accuracy due to the use of thin samples (transmission close to unity). This procedure was necessary in order to avoid large corrections due to preferential attenuation of neutrons near resonances thereby making it difficult to relate the average transmission to the true average cross section.

The computations by Lynn (1960b) of Doppler broadened area functions for wide ranges of resonance parameters and sample thickness enable the average transmission method to be used for thick samples. In addition, important effects due to interference between resonance and potential scattering are also included. Recently, Uttley and Jones (1961) have used both the average transmission and average cross section methods to determine the s- and p-wave strength functions of $(^{238}\text{U} + n)$ and $(^{232}\text{Th} + n)$. These authors point out that their analysis is inadequate as it is necessary to weight the area functions A_x with a Porter–Thomas distribution of reduced neutron widths (Porter and Thomas, 1956) when averaging over an energy interval which contains many resonances. Such a weighting factor is included by Lynn (1963a) in the analysis of the average $(^{238}\text{U} + n)$ and $(^{232}\text{Th} + n)$ data. The expression for the total cross section in the neighborhood of a single resonance (resonance energy E_0) may be obtained using the Thomas approximation given in Section I. It may be written (Lynn, 1963a):

$$\sigma_T(E) = \underset{\substack{\text{single}\\\text{level}\\\text{term}}}{\sigma_{\text{sl}}(E)} + \underset{\substack{\text{potential}\\\text{scattering}\\\text{term}}}{\sigma_{\text{pot}}} + \underset{\substack{\text{resonance–}\\\text{resonance}\\\text{interference}\\\text{term}}}{\sigma_{\text{rr}}} + \underset{\substack{\text{single level}\\\text{wings from}\\\text{other}\\\text{resonance}}}{\sigma_{\text{slw}}} \qquad (67)$$

where

$$\sigma_{\text{sl}}(x) = \sigma_0/(1 + x^2) + 2ka\sigma_0[x/(1 + x^2)]$$
$$\sigma_{\text{pot}} = 4\pi a^2(1 - R^{\infty\prime})^2$$
$$\sigma_{\text{rr}} \approx -4\pi a^2\pi^2 S_0{}^2 \qquad (\text{see Thomas, 1955})$$
$$\sigma_{\text{slw}} \approx -8\pi a^2(1 - R^{\infty\prime})\pi^2 S_0(\epsilon/3D)$$
$$+ 4\pi\lambda a S_0(\Gamma/D)\{(\pi^2/6) + (\pi^4/360) + \ldots\}$$

and

$$\epsilon = E - E_0$$

A uniform level approximation is assumed in which the spacing is D (eV), and the equal reduced widths are $\gamma_{\lambda n}^2$ so that $S_0 = \gamma_{\lambda n}^2/D$. The term σ_{rr} is essentially energy independent in the region of interest. The energy independent part of σ_{slw} may then be included with σ_{pot} and σ_{rr} to give an effective constant cross section σ_c, say. The neutron transmission is then given by

$$T(E) = \exp\{-n[\sigma_{\text{sl}}(E) + \sigma_c - 8\pi a^2(1 - R^{\infty\prime})\pi^2 S_0(\epsilon/3D)]\} \qquad (68)$$

For values of sample thickness n, used in practice, the final term in Eq. (68) may be expanded using the binominal theorem so that

$$T(E) \approx \exp\{-n[\sigma_{s1}(E) + \sigma_c]\}(1 + r\epsilon) \tag{69}$$

where

$$r = 8\pi na^2(1 - R^{\infty\prime})\pi^2(S_0/3D)$$

The average transmission $\bar{T}$ in the uniform level approximation is

$$\bar{T} = D^{-1} \int_{-D/2}^{+D/2} T(E) \, d\epsilon$$
$$= \exp(-n\sigma_c)\{D - \int_{-D/2}^{+D/2} [1 - \exp\{-n\sigma_{s1}(E)\}(1 + r\epsilon)] \, d\epsilon\} \tag{70}$$

where the integral in Eq. (70) is $\hat{A}_E(D/2)$ used by Lynn (1958).

The first order approximation to Eq. (70) is

$$\bar{T} = \exp\{-n\sigma_c\}[1 - (\Gamma/2D)A_x] \tag{71}$$

Lynn discusses this approximation in detail with special reference to the magnitude of the many level effects. He concludes that the higher order terms due to these effects are negligible in the cases of $(^{238}U + n)$ and $(^{232}Th + n)$ for the typical range of sample thicknesses used in experiments.

In order to obtain the mean transmission $\langle T \rangle$ of neutrons through many resonances, the average transmission $\bar{T}$ must be averaged over a Porter–Thomas distribution of neutron widths. We then have

$$\langle T \rangle = \exp\{-n\sigma_c\}[1 - G] \tag{72}$$

where

$$2DG = \int_0^\infty (\Gamma_\gamma + \Gamma_n)A_x P(\Gamma_n) \, d\Gamma_n$$

and

$$P(\Gamma_n) \, d\Gamma_n = (2\pi\Gamma_n\bar{\Gamma}_n)^{-1/2} \exp\{-\Gamma_n/2\bar{\Gamma}_n\} \, d\Gamma_n$$

Here, A_x is the Doppler broadened area function defined in Eq. (61).

If the self-screening effect of p-wave resonances is small (which is true in ^{239}U and ^{233}Th), the transmission of p-wave neutrons is given by $\exp\{-n\bar{\sigma}_p\}$ where $\bar{\sigma}_p$ is the average p-wave cross section. The average transmission for s- and p-wave neutrons is then

$$\langle T \rangle = \exp\{-n(\sigma_c + \bar{\sigma}_p)\}(1 - G) \tag{73}$$

The Thomas approach outlined in Section I is then used in its p-wave form to obtain $\bar{\sigma}_p$. Lynn gives

$$\bar{\sigma}_p = 6\pi\lambda^2(1 - \text{Re } \bar{U}_1) \tag{74}$$

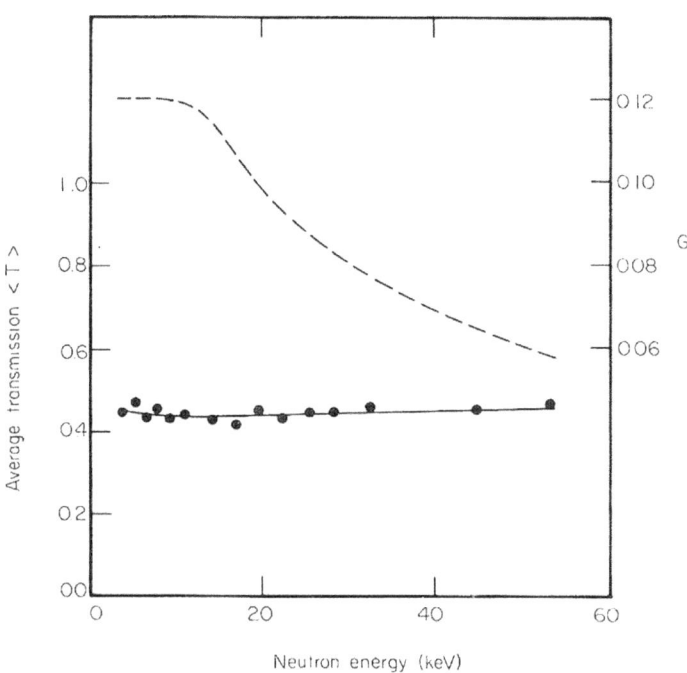

FIG. 14. An analysis of the average transmission data of ^{238}U in order to extract the p-wave strength function (Lynn, 1963a).

where the averaged collision function for p-waves is

$$\bar{U}_1 = \exp \{- 2i\phi_1\}(1 + iP_1\bar{R}_1)/(1 - iP_1\bar{R}_1) \qquad (75)$$

in which

$$\bar{R}_1 = R_1^\infty + i\pi S_1; \qquad S_1 = (\gamma_{\lambda n}^2/D)_{l=1}$$
$$P_1 = (ka)^3/[1 + (ka)^2]; \qquad \phi_1 = \tan^{-1}\{-j_1(ka)/n_1(ka)\}$$

Here, R_1^∞ is the cumulative effect of p-wave levels, and $j_1(ka)$ and $n_1(ka)$ are spherical Bessel and Neumann functions, respectively.

A least-squares fit to Eq. (73) is made to the data of (^{238}U $+ n$) and (^{232}Th $+ n$) using the s-wave potential scattering cross section (σ_{pot}) and the p-wave strength function (S_1) as variables. Lynn's results for ^{238}U are shown in Fig. 14. The accuracy obtained for the p-wave strength functions show clearly the power of this method compared with the determination of individual resonance parameters. (It is to be noted that more than 100 p-wave resonances would need to be identified and analyzed to obtain an error of $\pm 15\%$ on the p-wave strength function.)

IV. THE SELF-INDICATION METHOD

A. Experimental Technique

In the above sections, we have been discussing transmission measurements made with the use of a "flat" detector, i.e., one whose efficiency for detecting neutrons varies slowly with energy and is therefore constant over an energy interval in which the resonance effect of a single level is significant. Alternatively, it is possible to place the element under study at the detector position and observe, as a function of neutron energy, the γ-rays emitted in the process of radiative capture of the neutrons. There are two main motivations for this: (a) the desire to measure capture cross sections as a function of neutron energy in their own right, a topic discussed in detail in Chapter III; and (b) the fact that the response of the γ-ray detection system is very fast and the thickness of the primary detector, i.e., the sample, in the direction of the neutron flight path is negligible. Full advantage can therefore be taken of the resolution capability of the time-of-flight spectrometer system.

The second motivation was (until recently) the primary consideration in the operation of the Columbia University (Nevis) synchrocyclotron spectrometer system. The burst width and timing gates were both 0.1 μsec. In the detection system actually used, the sample under study was suspended normal to the beam direction, and the γ-rays emitted were detected by shielded plastic scintillation detectors placed above and below the sample.

Two distinct types of measurements can be made: one with various thicknesses of sample at the detector position only, and the other with the sample material at both the detector position and the transmission position (the "self-indication" method).

B. Analysis of the Data

1. SAMPLE AT DETECTOR POSITION ONLY

a. Data Are Taken for Various Thicknesses of Sample. If ϵ is the (unknown) efficiency for detecting γ-rays, a single resonance will then give a response in excess of background of $\epsilon(\Gamma_\gamma/\Gamma)(1 - T)$ (neutron flux). If a "thick" sample is included among the various samples, such that the transmission T is zero for several points in the neighborhood of the resonance energy, $\epsilon(\Gamma_\gamma/\Gamma)$ (neutron flux) is determined and can be used to convert the data for the various sample thicknesses to "$(1 - T)$ data." These data are then comparable with the transmission dips obtained with flat detectors and may be handled exactly as in Section III.

b. It Is Sometimes Observed That ϵ Is Constant for All the Levels of a Given Element. If the value of Γ_γ/Γ is known for one or more levels (obtained from other methods of measurements say), then a thick sample with $T = 0$ can be used to determine ϵ. (Occasionally, a simultaneous analysis of many levels can yield ϵ without recourse to external information.) For other levels one obtains $(\Gamma_\gamma/\Gamma)(1 - T)$. The area of the resonance can then be used as above. If the factor Γ_γ/Γ is not too close to unity, the areas can lead to a determination of the spin weighting factor g when these data are combined with conventional transmission measurements.

2. The Self-Indication Method: Samples at Both the Detector and Transmission Positions

Here, the element under investigation, together with the γ-ray detection system, is considered as a detector, and various thicknesses of the same element are used for making transmission measurements. Where the resolution warrants, channel-by-channel transmission values can be computed exactly as for flat detectors and the results treated similarly. Except at the lower neutron energies, this is seldom done because the resolution also effects the sample-out spectrum. The subsequent treatment of self-indication is based on area analysis. We define a quantity T_{si}, the self-indication transmission, as

$$T_{\mathrm{si}} = \frac{\text{area with transmission sample}}{\text{area without transmission sample}}$$

$$= \frac{\int_{-\infty}^{\infty} \{1 - \exp[-n_D\sigma_0\Psi(\beta,x)]\}\, \exp[-n_T\sigma_0\Psi(\beta,x)]\, dx}{\int_{-\infty}^{\infty} \{1 - \exp[-n_D\sigma_0\Psi(\beta,x)]\}\, dx} \tag{76}$$

$$= \frac{\int_{-\infty}^{\infty} \{1 - \exp[-(n_D + n_T)\sigma_0\Psi(\beta,x)]\}\, dx - \int_{-\infty}^{\infty} \{1 - \exp[-n_T\sigma_0\Psi(\beta,x)]\}\, dx}{\int_{-\infty}^{\infty} \{1 - \exp[-n_D\sigma_0\Psi(\beta,x)]\}\, dx} \tag{77}$$

where n_D and n_T refer, respectively, to the number of nuclei per square centimeter in the detector sample and in the transmission sample. The three integrals in the last expression for T_{si} are those referred to above as $A_x'(\infty,\beta)$. Hence, T_{si} can be calculated from previously evaluated expressions. Here, T_{si} is a function of $n_T\sigma_D$, $n_D\sigma_D$, and β, or any suitable combination of these. Graphs of these relationships have been prepared. As an example, Fig. 15 shows T_{si} versus $n\sigma_0\Gamma/\Delta$ for various values of Δ/Γ in the special case of $n_D = n_T$.

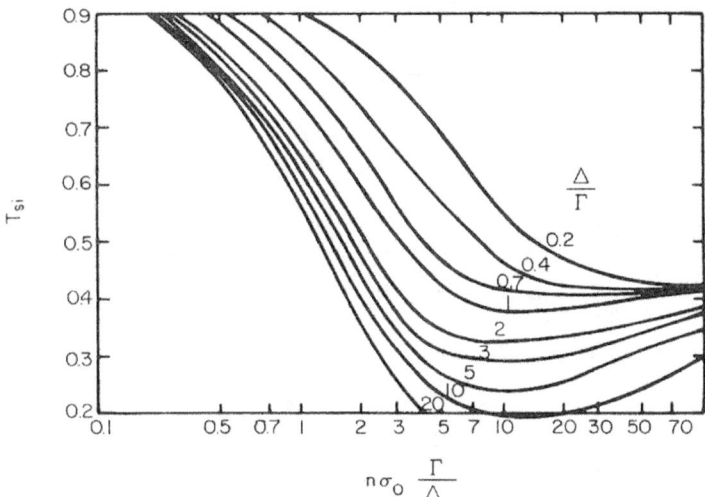

FIG 15. A set of curves of T_{si} versus $n\sigma_0\Gamma/\Delta$ for various values of Δ/Γ in the special case of $n_D = n_T$.

Each measurement of T_{si} with given values of n_D and n_T then gives a relationship between σ_0 and Γ (since n_D and n_T are known), and two measurements will then allow determinations of σ_0 and Γ. This, then, is the strict application of self-indication. Frequently, however, the curves of σ_0 versus Γ (or equivalent) for various measurements intersect at such small angles that small experimental errors result in a large uncertainty in the parameters. In this case, self-indication data may be combined with other measurements to deduce the parameters. The g value cannot, of course, be determined by self-indication alone.

Data on ^{238}U (Rosen et al., 1960) and on Ag, Au, and Ta (Desjardins et al., 1960) have been analyzed by a combination of methods, including self-indication. Figure 16 shows an example (Rosen et al., 1960) where self-indication on one transmission sample thickness is combined with an area measurement under the "D only" (capture yield) curve normalized by an internal determination of ϵ. Figure 17 (Desjardins et al., 1960) shows an example for Ag but with two different transmission sample thicknesses. Also, since $I = \frac{1}{2}$ for Ag, g is either $\frac{1}{4}$ or $\frac{3}{4}$, and the area under the "D only" curve gives a relationship for each value of g. Here, $J = 1$ ($g = \frac{3}{4}$) is clearly selected as the value for the spin J. Figure 18 shows an example for Au (Desjardins et al., 1960), with additional relationships deduced from flat detector measurements. These are examples of favorable cases in which

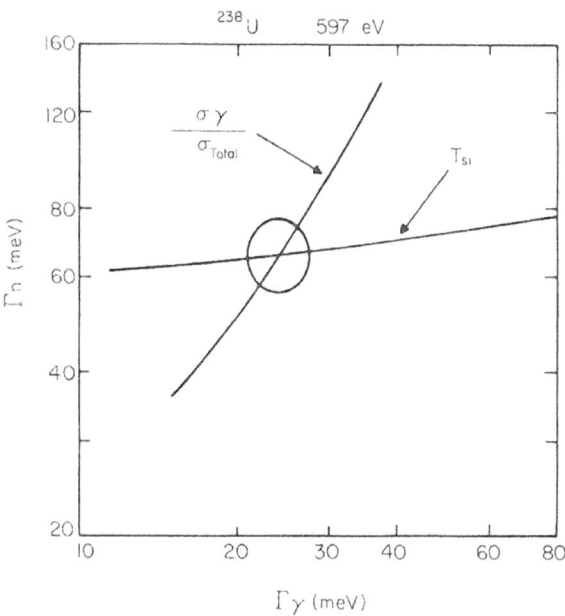

FIG. 16. Determination of resonance parameters using a combination of self-indication (transmission) data and capture yield data (Rosen *et al.*, 1960). ($\Gamma_n = 66 \pm 10$ meV; $\Gamma_\gamma = 24.0 \pm 3.0$ meV.)

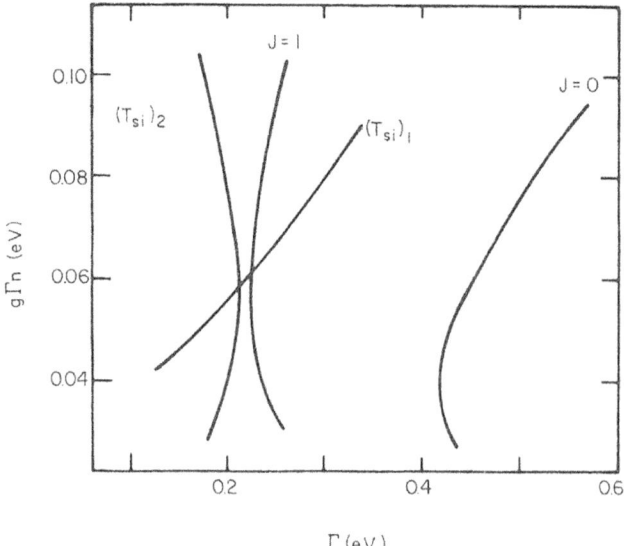

FIG. 17. Determination of resonance parameters using self-indication data with two different transmission sample thicknesses (Desjardins *et al.*, 1960). [Ag (134 eV).]

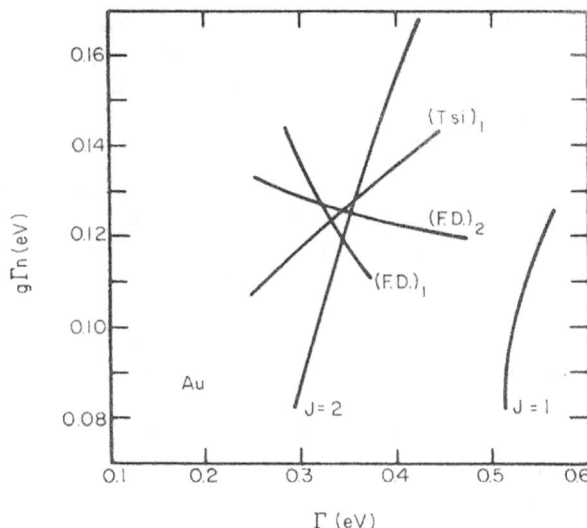

Fig. 18. Determination of resonance parameters using a combination of self-indication and conventional transmission data (Desjardins *et al.*, 1960). [Au (606 eV).]

complete evaluation of the parameters can be made. In many cases, only partial results can be obtained.

The measurements made by detection of γ-rays have some disadvantages when compared with transmission measurements, particularly for thicker samples: (i) the self-absorption of the capture γ-rays in the sample itself produces complicated variations in ϵ; (ii) scattering of neutrons of energy slightly higher than a resonance energy, but which now have a large probability of capture because their energy is reduced to the resonance energy; and (iii) the natural radioactivity of some samples. Corrections for multiple scattering of neutrons in the sample have been made recently by Lynn (1963b) using a Monte Carlo type calculation: such corrections greatly extend the scope of the self-indication method.

V. DATA OBTAINED FROM TOTAL NEUTRON CROSS SECTION MEASUREMENTS

Our present knowledge of neutron resonance parameters and the statistical properties of these parameters is based largely on the results of total cross section measurements. Although such measurements only give values of the resonance energy E_λ, the total width Γ_λ and the neutron width multiplied by the spin weighting factor $g\Gamma_\lambda$, a number of assumptions can frequently be made which enable the determination of other resonance parameters such as the radiation width Γ_γ or the total spin J. For example, a useful assumption is the near constancy of the total radiation width Γ_γ for a particular isotope. In the cases of medium and heavy elements ($A >$ 100), this assumption is theoretically well founded since there are many states which can be reached by γ-rays in the decay of a resonance to the ground state. Experimentally, this assumption has been established within an accuracy of better than 10% in a number of nuclei. A detailed discussion of both resonance parameters and their statistical properties is given by Lynn (1968). The following is an outline of the information which has been obtained from total cross section measurements.

A. Resonance Parameters

1. Measurement of the Total Radiation Width

At low neutron energies ($\gtrsim 100$ eV, say) radiative capture is the predominant process so that a measurement of the total width can yield accurate values of Γ_γ.

Cases of interest are:

(i) The neutron width extremely small so that $\Gamma_\gamma \approx \Gamma$. This straightforward case provided many of the early (and accurate) values of radiation widths (Sailor, 1953; Landon and Sailor, 1955).

(ii) The neutron width comparable with the radiation width ($\Gamma_n \approx \Gamma_\gamma$). In such cases, the radiation width is simply obtained from

$$\Gamma_\gamma = \Gamma - \Gamma_n \tag{78}$$

$$= \Gamma - \sigma_0 \Gamma / 4\pi \lambda^2 g \tag{79}$$

in which both σ_0 and Γ are known from accurate total cross section data. The spin gactor g presents no difficulty if the target spin I is zero ($g = 1$), or if I is large so that $g \approx \frac{1}{2}$. This method has been widely used to determine radiation widths.

2. Measurement of the Neutron Width

As mentioned above, the factor $g\Gamma_n$ is determined directly from total cross section measurements. For those elements in which $g = 1$ or $g \approx \frac{1}{2}$ a large number of neutron widths have been obtained (Harvey et al., 1955). In the case of ^{238}U $+ n$, 100 resonances have been analyzed and the values of Γ_n found (Firk et al., 1963a).

Other results with good statistical accuracy have been reported for the reaction $^{127}I + n$ (Garg et al., 1965) and the data found to agree with the Porter–Thomas distribution (see Section V,B,2).

3. MEASUREMENT OF THE RESONANCE SPINS

For resonances in the kilovolt region, the predominant decay process is by elastic neutron scattering in which case it is sometimes possible to determine the spin J of the resonances since

$$\sigma_0 \approx 4\pi\lambda^2 g(\Gamma_n/\Gamma) \approx 4\pi\lambda^2 g \qquad \text{for} \quad \Gamma_n \gg \Gamma_\gamma \tag{80}$$

An accurate measurement of the peak total cross section and the resonance energy determines the statistical weight factor g, and hence the spin J. This method has been used on a number of occasions when the resolution of the spectrometer has been sufficient to measure the peak cross section with precision (Bollinger *et al.*, 1955; Lynn *et al.*, 1958a; Good *et al.*, 1958; Firk *et al.*, 1963b; Coté *et al.*, 1964, and Morgenstern *et al.*, 1965).

B. Statistical Properties of Resonances

1. THE DISTRIBUTION OF LEVEL SPACINGS

The simplest distribution obtainable from slow neutron resonance data is the distribution of level spacings. Even in this case, however, a number of difficulties are soon encountered. For instance, the rapid decrease in resolving power of neutron spectrometers as the neutron energy increases results in an increasing number of unobserved resonances at high energy. In general, therefore, only a limited number of resonances at low energy are available for inclusion in the distributions. At the present time, the largest number of resonances clearly resolved in a single nuclide is about 200 (Garg *et al.*, 1964a).

Another frequent difficulty is due to lack of knowledge of the spins of resonances for those nuclei in which $I \neq 0$; for high spin values ($J > 3$, say), data are essentially nonexistent.

In spite of these limitations, the main features of the level spacing distributions are now known, however. It is clear from a number of experiments (Harvey and Hughes, 1958; Rosen *et al.*, 1960; Firk *et al.*, 1963a; and Garg *et al.*, 1964a) that the distribution is well described by that proposed by Wigner (1957), namely:

$$P(D)\,dD = D/2\bar{D}^2 \exp\{-D^2/4\bar{D}^2\}\,dD \tag{81}$$

where D is the level spacing and $\bar{D}$ is the mean level spacing. Equation (81) predicts zero probability of finding levels with zero spacing. This "level repulsion" effect had been noted by Gurevich and Pevzner (1956) and Lane *et al.* (1956).

The excellent agreement between the observations of Garg *et al.* (1964a) and the distribution given by Eq. (81) is demonstrated in Fig. 19.

50

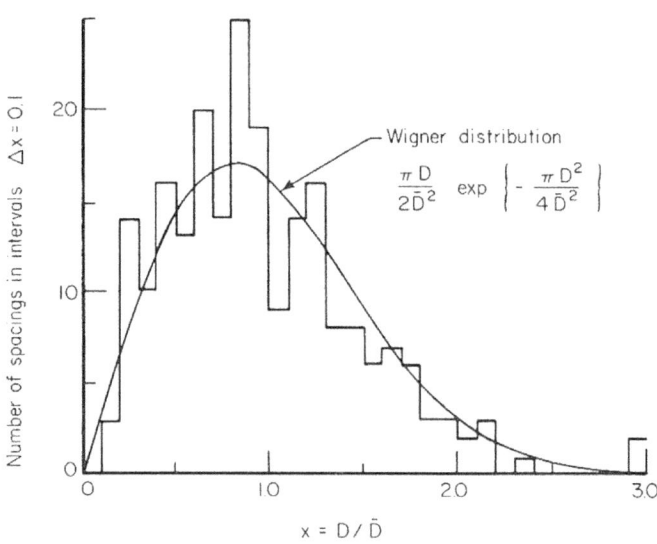

Fig. 19. The observed distribution of level spacings in $^{238}U + n$ compared with a Wigner distribution (Garg *et al.*, 1964a).

2. The Distribution of Reduced Neutron Widths

Using the data obtained by the Brookhaven fast chopper group prior to 1955, Hughes and Harvey (1955b) determined, empirically, the distribution of reduced neutron widths $[\Gamma_{\lambda n}^0 = \Gamma_{\lambda n}/(E_\lambda)^{1/2}]$. They concluded that the distribution has the form

$$P(x)\,dx = (2\sqrt{\pi})^{-1}(x/2)^{-1/2}\exp(-x/2)\,dx \qquad (82)$$

where

$$x = \Gamma_n^0/\langle\Gamma_n^0\rangle_{\mathrm{av}}$$

in which $\langle\Gamma_n^0\rangle_{\mathrm{av}}$ is the average reduced neutron width.

This distribution was predicted theoretically by Brink (1955) who argued that the reduced width amplitudes [see Eq. (18)] should have a Gaussian distribution with zero mean, thus

$$P(\gamma_{\lambda n})\,d\gamma_{\lambda n} = (2\pi\langle\gamma_{\lambda n}^2\rangle_{\mathrm{av}})^{-1/2}\exp(-\gamma_{\lambda n}^2/2\langle\gamma_{\lambda n}^2\rangle_{\mathrm{av}})\,d\gamma_{\lambda n} \qquad (83)$$

so that the distribution of reduced widths becomes

$$P(\gamma_{\lambda n}^2)\,d\gamma_{\lambda n}^2 = (2\pi\gamma_{\lambda n}^2\langle\gamma_{\lambda n}^2\rangle_{\mathrm{av}})^{-1/2}\exp(-\gamma_{\lambda n}^2/2\langle\gamma_{\lambda n}^2\rangle_{\mathrm{av}})\,d\gamma_{n\lambda}^2 \qquad (84)$$

Porter and Thomas (1956) derived Eq. (84), independently, using arguments based upon the physical properties of the overlap integral in the

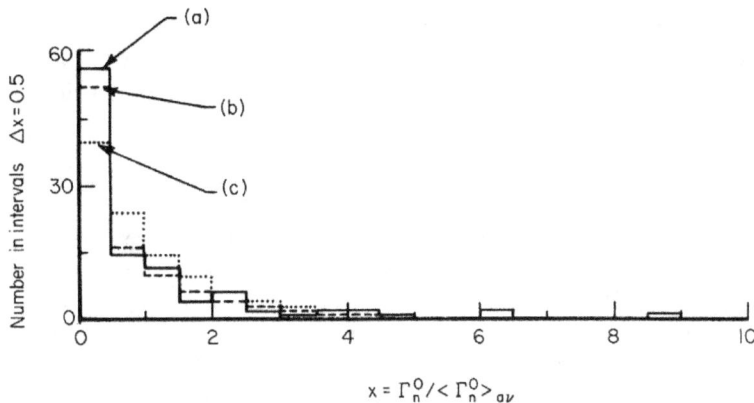

FIG. 20. The observed distribution of reduced neutron widths for 100 levels in ^{238}U + n compared with Porter–Thomas and exponential distributions (Firk *et al.*, 1963a). (a) Experimental data; (b) theoretical fit: a Porter–Thomas distribution; (c) theoretical fit: an exponential distribution.

definition of the reduced width amplitude [see Eq. (18)]. They concluded that positive and negative values of the integrand are equally likely which is equivalent to Brink's conjecture.

The Porter–Thomas–Brink frequency function is a special case ($\nu = 1$) of the chi-squared frequency function with ν degrees of freedom:

$$\rho_\nu(x)\ dx = \bar{\Gamma}^{-1}(\nu/2)(\nu/2\bar{x})^{\nu/2}x^{(\nu-2)/2}e^{-\nu x/2\bar{x}}\ dx \qquad (85)$$

where $\bar{\Gamma}$ is the incomplete gamma function.

An exponential frequency function is obtained from Eq. (85) for a value of $\nu = 2$.

In order to find the frequency function which best fits the data, it is necessary to determine ν and the error associated with it. The maximum likelihood method is used (Kendall, 1946). For the chi-squared family the most likely value of ν is given by

$$n^{-1}\sum_{i=1}^{n}\ln x_i = \ln \bar{x} = \phi(\nu/2) - \ln(\nu/2) \qquad (86)$$

where x_i ($i = 1$ to n) are the observed values, $\bar{x}$ is the mean value, and $\phi(\nu/2)$ is the logarithmic derivative of $\bar{\Gamma}(\nu/2)$.

The most recent results based upon 416 reduced neutron widths (Desjardins *et al.*, 1960; Firk *et al.*, 1963a; Garg *et al.*, 1964a; and Garg *et al.*, 1965) have been analyzed by Garrison (1964), and a value of $\nu = 1.04 \pm 0.10$ obtained which is in excellent agreement with the distribution given in Eq. (84). A typical example is shown in Fig. 20 in which the observed distribution of 100 reduced neutron widths in ^{238}U $+ n$ (Firk et al., 1963a) is shown together with the theoretical distributions for $\nu = 1$ and $\nu = 2$.

C. Neutron Strength Functions

One of the most significant quantities obtained from the results of slow neutron total cross section studies is the s-wave neutron strength function and its variation with mass number. This quantity is directly related to the average compound nucleus cross section (Feshbach et al., 1954). For example, on integrating over a Breit–Wigner line shape (ignoring interference and resonant self-absorption effects) and then summing over an appropriate energy interval E', we find

$$\sigma_{\text{res}} \times E' = (4\pi^2/k^2)\rho \sum_{\lambda} \gamma_{\lambda n}^2 \tag{87}$$

where $\rho = ka$, and k is assumed to be constant throughout E'. If the average spacing between levels is $\bar{D}$, and there are λ levels in E', and $\langle \gamma_n^2 \rangle_{\text{av}}$ is the average reduced neutron width, then

$$\bar{\sigma}_{res} = (4\pi^2/k^2)\rho S_0 \tag{88}$$

where

$$S_0 = \langle \gamma_n^2 \rangle_{\text{av}}/\bar{D}$$

is the s-wave neutron strength function.

In the theoretical development of the complex potential model of nuclear reactions, the R-function defined in Eq. (17) for elastic scattering becomes modified

$$R_{\text{op}} = \sum_P \gamma_P^2/(E_P - E - iW) \tag{89}$$

where a complex potential $V = -(V_0 + iW)$ has been used in the original Schrödinger Eq. (3), and E_P is the energy of a single particle resonance in the real potential V_0.

The strength function S is related to the imaginary part of R_{op} (Lane and Thomas, 1958), thus

$$R_{\text{op}} = R^{\infty} + i\pi S \tag{90}$$

so that

$$S = \frac{W}{\pi} \sum_P \frac{\gamma_P^2}{(E_P - E)^2 + (W)^2} \tag{91}$$

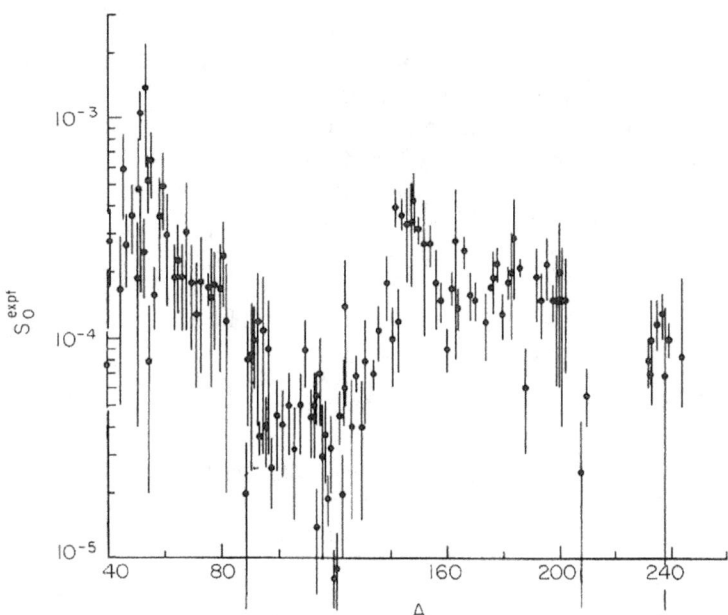

FIG. 21. The observed s-wave neutron strength function S_0^{expt} as a function of mass number (Lynn, 1968).

It is not practicable to study the location of these single particle resonances in a particular nucleus because many l-waves participate at higher energies so that the underlying simple structure becomes ill defined. An alternative approach is to study the variations of the strength function, normalized to an energy of 1 eV, as a function of mass number A (equivalent to a variation in nuclear radius). The advantage of this method is due to the dominance of s-wave interactions at such a low energy, thus enabling the determination of accurate values of neutron widths for many resonances whose spins are known. It is therefore customary to define the experimental neutron strength function for s-wave as

$$S_0^{\text{expt}} = \langle \Gamma_{\lambda n}^0 \rangle_{av} / \bar{D} = 2k_1 a S_0 \tag{92}$$

where

$$\Gamma_{\lambda n}^0 = \Gamma_{\lambda n} / (E_\lambda/1)^{1/2}$$

and k_1 is the neutron wave number at 1 eV.

The most recent compilation of data giving S_0^{expt} as a function of A is due to Lynn (1968) and is shown in Fig. 21.

The general form of the curve agrees well with the predictions of the

53

original complex potential model of Feshbach *et al.* (1954) who used a square complex potential with a radius constant $r_0 = 1.45 \times 10^{-13}$ cm and a depth for the real part of the well $V_0 = -42$ MeV. Maxima occur in the strength function at $A \sim 60$ (3s-state) and at $A \sim 155$ (4s-state). The early experimental work of Carter *et al.* (1954) resulted in a best fit to the data using a value of $W_0 = -3.4$ MeV for the depth of the imaginary part of the potential.

Many refinements to the theory have since been reported in order to obtain improved agreement with experiment. The most important features of these refinements are:

(i) the use of a potential well with a diffuse edge (e.g., Woods–Saxon form);

(ii) the use of a surface peaked imaginary part of the potential;

(iii) the use of a nonspherical complex potential (static); and

(iv) inclusion of rotational and vibrational modes of the target nucleus induced by the incident nucleon.

Effects such as the possible coupling of the first 2^+ excited state to the two-phonon triplet (0^+, 2^+, 4^+) have been considered by Furuoya and Sugie (1963), but the predicted third peak associated with the 3s-state is not evident in the present results.

Possible effects due to spin-orbit splitting, particularly for the d-state which is important in studying the coupling of rotations and quadrupole vibrations, are likewise not evident. However, the accuracy of many strength function measurements is not sufficiently high to rule out the possibility of these more subtle effects.

Since the existence of spin exchange forces in the nucleon-nucleon interaction is established, it is not unreasonable to postulate the existence of a spin-spin term in the optical potential which couples the target spin I to the incident particle spin i in a manner analogous to the isospin coupling term. Such a term would result in an angular momentum dependence of the s-wave neutron strength function (Firk *et al.*, 1963b). Recent evidence of such effects has been reported in ^{77}Se $+ n$ (Julien *et al.*, 1962), in ^{75}As $+ n$ (Julien *et al.*, 1964; Garg *et al.*, 1964b) in ^{59}Co $+ n$ (Morgenstern *et al.*, 1965) and in ^{197}Au $+ n$ by Julien *et al.* (1966).

Perhaps a clearer demonstration of spin–spin effects is obtained from the observed spin dependence of the scattering length a_J [see Eq. (47)] deduced in the case of ^{51}V $+ n$ by Firk *et al.* (1963b). Lynn (1968) estimates that the difference in scattering lengths in ^{51}V $+ n$ leads to a value of the coupling strength $V_2 \sim 10$ W where

$$\Delta V_{i \cdot I} = (V_2/A)(\boldsymbol{i} \cdot \boldsymbol{I}) \tag{93}$$

In addition to the *s*-wave strength function, several measurements of the *p*-wave strength function have been reported. The two main methods used to deduce the *p*-wave strength function are:

(i) deviations (for small reduced neutron widths) from the *s*-wave Porter–Thomas distribution (Rosen *et al.*, 1960; Desjardins *et al.*, 1960; Michaudon and Ribon, 1962; Garg *et al.*, 1965; Le Poittevin *et al.*, 1965); and

(ii) analysis of average transmissions in the kilovolt region (Uttley and Jones, 1961; Lynn, 1963a; and Newstead, 1967).

Although both these methods have many difficulties associated with them, it is possible to obtain *p*-wave strength functions with errors as low as $\pm 15\%$ in favorable cases.

At the present time, there is no conclusive evidence to support the contention that the *p*-wave strength function exhibits a spin–orbit splitting in the region $A \sim 100$.

It is clear from the above results that total cross section measurements form the backbone of slow neutron spectroscopy, and that they will continue to do so in the future. The performance of modern electron and proton accelerators indicates that time-of-flight resolutions of less than 0.01 nsec m^{-1} will shortly be achieved in the energy range above several kilovolts. The majority of resonances in all nuclei will then be resolved up to an energy of about 10 keV. The analysis of the data from total, capture, and scattering cross section measurements (Rae *et al.*, 1958; Evans *et al.*, 1959; Iliescu *et al.*, 1965; and Asghar *et al.*, 1966) will give greatly improved values of both the individual and the statistical properties of neutron resonances, thereby adding to our general knowledge of nuclear structure.

REFERENCES

ASGHAR, M., CHAFFEY, C. M., MOXON, M. C., PATTENDEN, N. J., RAE, E. R., and UTTLEY, C. A. (1966). *Nucl. Phys.* **76**.

BETHE, H. A., and PLACZEK, G. (1937). *Phys. Rev.* **51**, 462.

BOLLINGER, L. M., DAHLBERG, D. A., PALMER, R. R., and THOMAS, G. E. (1955). *Phys. Rev.* **100**, 126.

BOLLINGER, L. M., THOMAS, G. E., and GINTHER, R. J. (1959). *Rev. Sci. Instr.* **30**, L.1135.

BOLLINGER, L. M., THOMAS, G. E., and GINTHER, R. J. (1962). *Nucl. Instr. Methods* **17**, 97.

BREIT, G. (1940). *Phys. Rev.* **58**, 506.

BREIT, G. (1946). *Phys. Rev.* **69**, 472.

BREIT, G., and WIGNER, E. P. (1936). *Phys. Rev.* **49**, 519, 642.

BRINK, D. M. (1955). Thesis, Univ. of Oxford (unpublished).

BROOKS, F. D. (1961). *In* "Neutron Time-of-Flight Methods" (J. Spaepen, ed.), p. 389. E.A.E.C. (Euratom), Brussels.

CARTER, R. S., HARVEY, J. A., HUGHES, D. J., and PILCHER, V. E. (1954). *Phys. Rev.* **96**, 113.

CARPENTER, R. T., and BOLLINGER, L. M. (1960). *Nucl. Phys.* **21**, 66.

CORGE, C., HUYNH, V-D., JULIEN, J., MORGENSTERN, J., and NETTER, F. (1961). *In* "Neutron Time-of-Flight Methods" (J. Spaepen, ed.), p. 545. E.A.E.C. (Euratom), Brussels.

COTÉ, R. E., BOLLINGER, L. M., and THOMAS, G. E. (1964). *Phys. Rev.* **134**, B1047.

DESJARDINS, J. S., ROSEN, J. L., HAVENS, W. W., JR., and RAINWATER, J. (1960). *Phys. Rev.* **120**, 2214.

EVANS, J. E., KINSEY, B. B., WATERS, J. R., and WILLIAMS, G. H. (1959). *Nucl. Phys.* **9**, 205.

FESHBACH, H., PORTER, C. E., and WEISSKOPF, V. F. (1954). *Phys. Rev.* **96**, 448.

FIRK, F. W. K. (1958). *Nucl. Phys.* **9**, 198.

FIRK, F. W. K., SLAUGHTER, G. G., and GINTHER, R. J. (1961). *Nucl. Instr. Methods.* **13**, 313.

FIRK, F. W. K., LYNN, J. E., and MOXON, M. C. (1963a). *Nucl. Phys.* **41**, 614.

FIRK, F. W. K., LYNN, J. E., and MOXON, M. C. (1963b). *Proc. Phys. Soc.* **82**, 201.

FURUOYA, I., and SUGIE, A. (1963). *Nucl. Phys.* **44**, 44.

GARG, J. B., RAINWATER, J., PETERSON, J. S., and HAVENS, W. W., JR. (1964a). *Phys. Rev.* **134**, B985.

GARG, J. B., HAVENS, W. W., JR., and RAINWATER, J. (1964b). *Phys. Rev.* **136**, B177.

GARG, J. B., RAINWATER, J., and HAVENS, W. W., JR. (1965). *Phys. Rev.* **137**, B547.

GARRISON, J. D. (1964). *Ann. Phys.* **30**, 269.

GAYTHER, D. B., and NICHOLSON, K. P. (1957). *Proc. Phys. Soc.* **A 70**, 51.

GOOD, W. M., NEILER, J. H., and GIBBONS, J. H. (1958). *Phys. Rev.* **109**, 926.

GUREVICH, I. I., and PEVZNER, M. I. (1956). *Proc. Intern. Conf. Nucl. Reactions, Amsterdam. Physica* **XXII**, 1132.

HARVEY, J. A., and HUGHES, D. J. (1958). *Phys. Rev.* **109**, 471.

HARVEY, J. A., and ATTA, S. E. (1961). *In* "Neutron Time-of-Flight Methods" (J. Spaepen, ed.), p. 55. E.A.E.C. (Euratom), Brussels.

HARVEY, J. A., HUGHES, D. J., CARTER, R. E., and PILCHER, V. E. (1955). *Phys. Rev.* **99**, 10.

HAVENS, W. W., JR., and RAINWATER, L. J. (1946). *Phys. Rev.* **70**, 154.

HAVENS, W. W., JR., and RAINWATER, L. J. (1951). *Phys. Rev.* **83**, 1123.

HUGHES, D. J. (1955a). *J. Nucl. Energy* **1**, 237.

HUGHES, D. J., and HARVEY, J. A. (1955b). *Phys. Rev.* **99**, 1032.

HUGHES, D. J., and ZIMMERMAN, R. L. (1956). *In* "Nuclear Reactions" (P. M. Endt and P. B. Smith, eds.), Vol. 1, p. 380. North Holland Publ., Amsterdam.

HUGHES, D. J., and SCHWARTZ, R. B. (1958). Brookhaven National Laboratory Rept., BNL 325, 2nd ed.

HUMBLET, J., and ROSENFELD, L. (1961). *Nucl. Phys.* **26**, 529.

ILIESCU, N., SAN, KIM, H., PIKELNER, L. B., SHARAPOV, E. I., and SIRAZHET, H. (1965). *Nucl. Phys.* **72**, 298.

JACKSON, H. E., and LYNN, J. E. (1962). *Phys. Rev.* **127**, 461.

JULIEN, J., CORGE, C., HUYNH, V. D., MORGENSTERN, J., and NETTER, F. (1962). *Phys. Letters* **3**, 69.

II. TOTAL NEUTRON CROSS SECTION MEASUREMENTS   153

JULIEN, J., BIANCHI, G., CORGE, C., HUYNH, V. D., LE POITTEVIN, G., MORGENSTERN, J., NETTER, F., and SAMOUR, C., (1964). *Phys. Letters* **10**, 86.

JULIEN, J., DE BARROS, S., BIANCHI, G., CORGE, C., HUYNH, V. D., LE POITTEVIN, G., MORGENSTERN, J., NETTER, F., SAMOUR, C., and VASTEL, M. (1966). *Nucl. Phys.* **76**, 391.

KAPUR, P. L., and PEIERLS, R. (1938). *Proc. Roy. Soc. (London)* **A166**, 277.

KENDALL, M. G. (1946). "The Advanced Theory of Statistics," Vol. II, Chapter 17. Griffin, London.

KHAN, F. A., and HARVEY, J. A. (1962). *Bull. Am. Phys. Soc.* **7**, 289.

KROTKOV, R. (1955). *Can. J. Phys.* **33**, 622.

LADENBERG, R., and REICHE, F. (1913). *Ann. Phys.* **42**, 181.

LAMB, W. E. (1939). *Phys. Rev.* **55**, 190.

LANDON, H. H. (1954). *Phys. Rev.* **94**, 1215.

LANDON, H. H., and SAILOR, V. L. (1955). *Phys. Rev.* **98**, 1267.

LANE, A. M., and THOMAS, R. G. (1958). *Rev. Mod. Phys.* **30**, 257.

LANE, A. M., LYNN, J. E., and STORY, J. S. (1956). Atomic Energy Research Establishment, Harwell, England, Rept. No. T/M 137.

LE POITTEVIN, G., DE BARROS, S., HUYNH, V. D., JULIEN, J., MORGENSTERN, J., NETTER F., and SAMOUR, C. (1965). *Nucl. Phys.* **70**, 497.

LYNN, J. E. (1958), *Nucl. Phys.* **7**, 599.

LYNN, J. E. (1960a). *Nucl. Instr. Methods* **9**, 315.

LYNN, J. E. (1960b). Atomic Energy Research Establishment, Harwell, England, Rept. Nos. R-3353 and R-3354.

LYNN, J. E. (1963a). *Proc. Phys. Soc.* **82**, 903.

LYNN, J. E. (1963b). Private communication.

LYNN, J. E. (1968). "The Theory of Neutron Resonance Reactions." Univ. Press, Oxford.

LYNN, J. E., and RAE, E. R. (1957). *J. Nucl. Energy* **4**, 418.

LYNN, J. E., FIRK, F. W. K., and MOXON, M. C. (1958a). *Nucl. Phys.* **5**, 603.

LYNN, J. E., MOXON, M. C., and FIRK, F. W. K. (1958b). *Nucl. Phys.* **7**, 613.

McDANIEL, B. D. (1946). *Phys. Rev.* **70**, 832.

MELKONIAN, E. (1956). *Proc. Intern. Conf. Peaceful Uses At. Energy, Geneva* **4**.

MELKONIAN, E., HAVENS, W. W., JR., and RAINWATER, J. (1953). *Phys. Rev.* **92**, 702.

MEYER, R. R. (1949). *Phys. Rev.* **75**, 773.

MICHAUDON, A. (1961). *In* "Neutron Time-of-Flight Methods" (J. Spaepen, ed.), p. 531. E.A.E.C. (Euratom), Brussels.

MICHAUDON, A., and RIBON, P. (1962). Private communication.

MORGENSTERN, J., BIANCHI, G., CORGE, C., HUYNH, V. D., JULIEN, J., NETTER, F., LE POITTEVIN, G., and VASTEL, R. (1965). *Nucl. Phys.* **62**, 529.

MÖSSBAUER, R. L. (1958). *Z. Physik.* **151**, 124.

MUEHLHAUSE, C. O., and THOMAS, G. E. (1953). *Nucleonics* **11**, 44.

NEWSTEAD, C. M. (1967). Thesis, Univ. of Oxford (unpublished).

PETERSON, S. W., and LEVY, H. A. (1952). *Phys. Rev.* **87**, 462.

PORTER, C. E., and THOMAS, R. G. (1956). *Phys. Rev.* **104**, 483.

RAE, E. R., and BOWEY, E. M. (1953). *Proc. Phys. Soc.* **A66**, 1073.

RAE, E. R., COLLINS, E. R., KINSEY, B. B., LYNN, J. E., and WIBLIN, E. R. (1958). *Nucl. Phys.* **5**, 89.

RIBON, P., DIMITRIJEVICK, Z., MICHAUDON, A., and WAGNER, P. (1961). *J. Phys. Radium* **22**, 708.

ROSE, M. E., MIRANKER, W., LEAK, P., and RABINOWITZ, G. (1953). Brookhaven Natl. Lab. Rept, BNL. 257.

ROSEN, J. L., DESJARDINS, J. S., RAINWATER, J., and HAVENS, W. W. JR., (1960). *Phys. Rev.* **118,** 687.

SAILOR, V. L. (1953). *Phys. Rev.* **91,** 53.

SEIDL, F. G. P., HUGHES, D. J., PALEVSKY, H., LEVIN, J. S., KATO, W. Y., and SJÖSTRAND, N. G. (1954). *Phys. Rev.* **95,** 476.

SETH, K. K. (1959). *Ann. Phy.* **8,** 223.

STEHN, J. R., GOLDBERG, M. D., WIENER-CHASMAN, RENATE, MUGHABGHAB, S. F., MAGURNO, B. A., and MAY, V. M. (1965). Brookhaven Natl. Lab. Rept., BNL 325, 2nd ed., supplement 2.

THOMAS, R. G. (1955). *Phys. Rev.* **97,** 224.

UTTLEY, C. A., and JONES, R. H. (1961). *In* "Neutron Time-of-Flight Methods" (J. Spaepen, ed.), p. 109. E.A.E.C. (Euratom), Brussels.

UTTLEY, C. A. and JONES, R. H. (1962). Private communication.

VOGT, E. (1959). *In* "Nuclear Reactions" (P. M. Endt and M. Demeur, eds.), Vol. 1, p. 215. North-Holland Publ., Amsterdam.

VOGT, E. (1962). *Rev. Mod. Phys.* **34,** 723.

VOITOVETSKII, V. K., TOLMACHEVA, N. S., and ARSAEV, M. I. (1959). *At. Eneng.* **6,** 321, 472.

WIGNER, E. P. and EISENBUD, L. (1947). *Phys. Rev.* **72,** 29.

WIGNER, E. P. (1957). *Proc. Gatlinburg Conf. Neutron Time-of-Flight Methods*, Oak Ridge Natl. Lab. Rept., O.R.N.L. 2309.

APPENDIX II

Recent Developments in Neutron Detection

F. W. K. FIRK*

Atomic Energy Research Establishment, Harwell, England

1. Introduction

Neutron detectors with increased efficiency, reduced background, and faster time resolution are constantly being sought. Two advances have been reported recently which provide neutron detectors with some of the characteristics mentioned above. First, the pulse-shape discrimination method, pioneered by Brooks (Br56,Br58, Br59,Br60), makes possible the efficient detection of fast neutrons when using certain organic scintillators in the presence of high γ-ray backgrounds. Second, the development by Ginther and Schulman (Gi58,Gi60) of glass scintillators, loaded with either boron or lithium, provides neutron detectors of reasonable efficiency and fast time resolution. The use of pulse-shape discrimination has rapidly become a standard technique in fast neutron spectroscopy. Experiments using Li^6-loaded glass scintillators (Fi61,Sl61) have recently demonstrated the usefulness of these scintillators for detecting neutrons by the nanosecond time-of-flight technique in the difficult energy region from 10 to 100 kev.

2. The Pulse-Shape Discrimination Method

A. *Decay Times and Light Intensities*

In 1956 Wright (Wr56) reported that the scintillations produced by α particles and electrons in anthracene crystals had different decay times, 53 and 31 nanoseconds, respectively. Wright assumed in his analysis that the scintillation pulse could be described by an initial

---

* This work was done while the author was assigned as an exchange visitor with the Oak Ridge National Laboratory.

fast spike (duration $\gtrsim$ 10 nanoseconds), followed by a longer exponentially decaying component. The figures of 53 and 31 nanoseconds refer to the latter component. Shortly afterwards Brooks (Br56,Br56a) developed a scintillation counter and discrimination circuit capable of distinguishing fast neutrons from γ rays by means of the different "effective decay times" of the recoil proton and Compton electron scintillations. The first detector used by Brooks

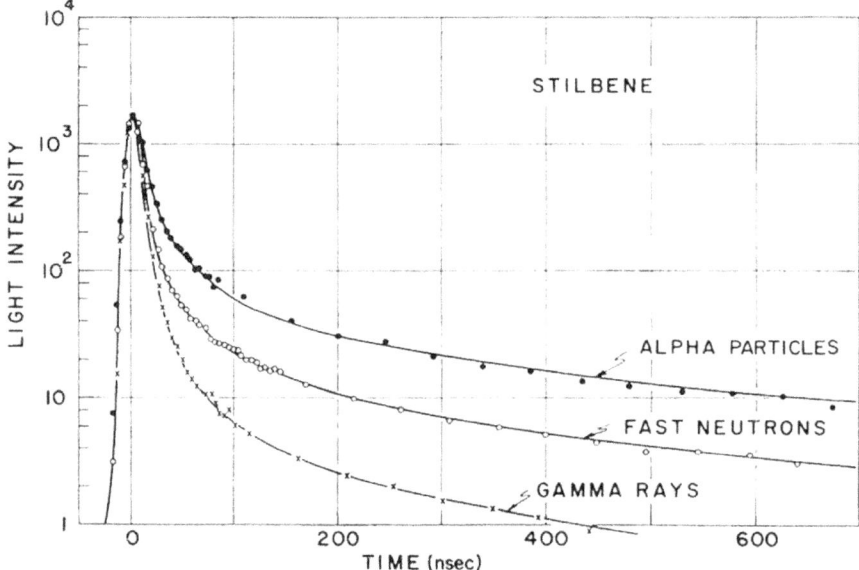

Figure 1. The scintillation response of stilbene to α particles, protons, and electrons normalized to the same peak height (Bo61).

(Br56a), incorporating a 1-in. thick stilbene crystal, detected 2-Mev neutrons with an efficiency of 10 per cent, whereas the detection efficiency for 2-Mev γ rays was less than 0.01 per cent.

The use of the term "effective decay time" is necessary to describe the pulse-shape discrimination system originally used by Brooks, since the work of Owen (Ow58) showed that the *decay times* of both the initial fast spike and the long-lived components, observed in organic scintillators, are in fact *independent* of the type of exciting particle. The differences occur in the *intensity* of the long-lived com-

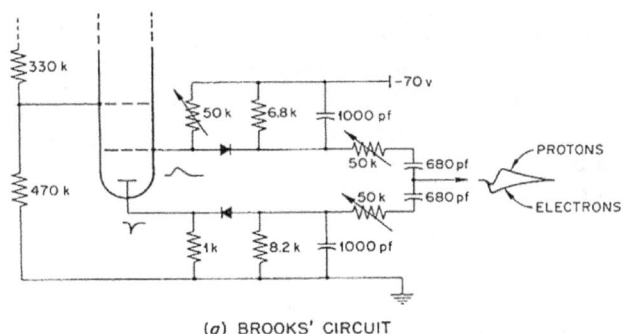

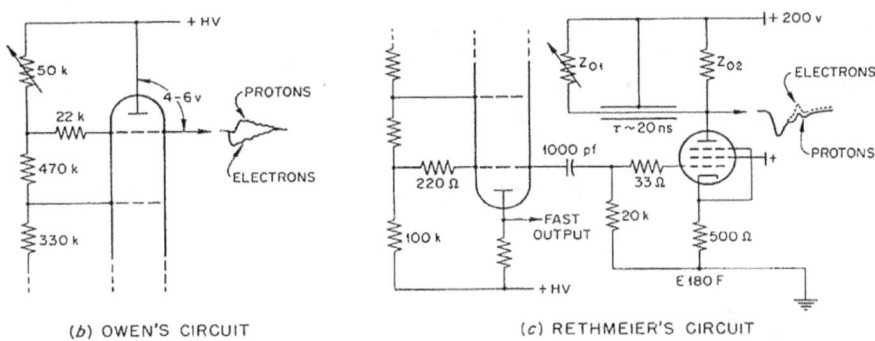

Figure 2. The three categories of pulse-shape discrimination circuits currently in use.

ponents relative to the intensity of the initial fast spike.[1] A typical example is shown in Fig. 1 taken from a recent paper by Bollinger and Thomas (Bo61). They have made precise measurements of the scintillation pulse shapes, as a function of the type of exciting particle, for a variety of organic and inorganic scintillators. Referring to Fig. 1, the following inequalities hold:

$$R_{\text{alpha}} < R_{\text{proton}} < R_{\text{electron}}$$

where

$$R = \int_0^\tau I(t)\,dt \bigg/ \int_\tau^\infty I(t)\,dt, \quad \tau \approx 10 \text{ nanoseconds} \tag{1}$$

[1] Neiler and Pethe (Ne61) have recently demonstrated that the frequency spectrum of light emitted in stilbene and anthracene crystals differs for α-particle or electron excitation. The intensity of ultraviolet emission is enhanced in the α-particle case.

Many circuits have been reported during the past few years which are sensitive to the ratio of electric charge given by Eq. (1) (Br58, Fo58,Ow59,Br59,Ba60,Bl60,Br60,Da61,Re61). It is usual to use a simple voltage discriminator to simulate the inequality sign. The various circuits fall into three categories, shown in Fig. 2. Each has certain features worthy of attention.

B. Circuit Details

(1) **The Brooks Circuit.** This circuit compares the total light output during the scintillation with the light output during the initial fast spike. A relatively long integrating time constant is required (1 to 2 μsec) to obtain the total light signal. There are both advantages and disadvantages associated with this long time constant. The advantages arise from the fact that good discrimination is obtained for recoil proton energies as low as 200 kev where it is essential

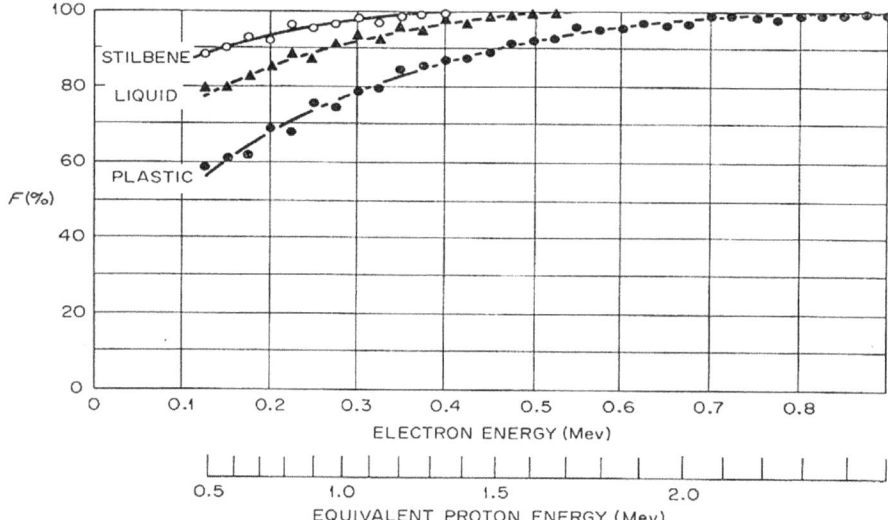

Figure 3. The quantitative performance of a typical Brooks' circuit using (a) a stilbene crystal; (b) a liquid scintillator (Nuclear Enterprises, Ltd., Type NE213); and (c) a plastic scintillator Type NE77. A bias δ is set to eliminate > 99 per cent of electron scintillations of any energy. The ordinate gives the percentage, F, of proton scintillations counted above δ, as a function of proton energy.

APPENDIX II   2241

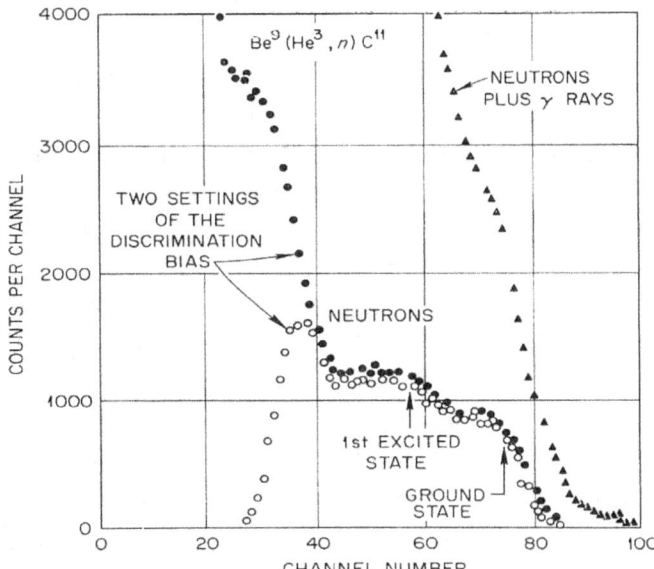

Figure 4. Pulse-height spectra from a liquid scintillator using the Be⁹(He³,n)C¹¹ reaction demonstrating the effect of an Owen pulse-shape discrimination circuit (Li59).

to collect as much information from the long-lived (but low intensity) components as possible. The disadvantages arise from a number of sources. The dead time of several microseconds can be a serious problem where high background counting rates are encountered. Also, the use of nonlinear circuit elements and two independent output terminals (the last dynode and collector) introduces difficulties due to gain shifts. In spite of these limitations the circuit has been used with success (Br59a). Figure 3 gives quantitative data on the performance of a typical Brooks circuit using stilbene, liquid, and plastic scintillators. (The problems associated with using two electrodes have been eliminated by Bloom *et al.* (Bl60), whose method requires a signal from a single electrode. The discrimination circuit is, however, still of the Brooks variety.)

(2) **The Owen Circuit.** Owen (Ow59) has approached the problem in a unique and simple way. By operating the last dynode and collector at a potential difference of only a few volts, it is possible to obtain space charge saturation of the last dynode during the initial

(and intense) fast spike of the scintillation. The output from the last dynode is a negative pulse during this short period of time. During the long-lived (and low intensity) component there is no space charge saturation. A suitable adjustment of the circuit parameters in Fig. 2(b) can be made to yield a positive signal at the last dynode from proton scintillations and a negative signal from electron scintillations [since $R_{\text{proton}} < R_{\text{electron}}$ in Eq. (1)]. The circuit suffers somewhat by requiring reasonably intense signals to operate satisfactorily. However, for detecting neutrons above an energy of $\sim$500 kev this is a popular system. A typical application is demonstrated in Fig. 4, taken from a paper by Litherland et $al.$ (Li59).

(3) **The Rethmeier Circuit.** A technique of considerable interest has recently been reported by Rethmeier, Boersma, and Jonker (Re61). This employs the use of a single electrode and the use of passive circuit elements only. Referring to Fig. 2(c), it is seen that suitable mismatch of the delay line in the anode of the E180F amplifier results in a reflected pulse which can be adjusted to be above a certain discrimination voltage for electron scintillations and below for proton scintillations. Apart from the advantages of using a single electrode and passive circuit elements, the circuit is designed to provide a fast timing signal from the collector of the phototube. The delay time of the reflected signal is only 40 nanoseconds, so that high counting rates should be readily handled by the system. The major disadvantage would appear to be a lack of discrimination for proton energies less than 800 kev.

It is clear from the many experiments reported recently in the field of fast neutron physics that the identification of the type of particle by means of pulse-shape discrimination is one of the most powerful techniques to emerge in recent years.

3. Glass Scintillators

The difficulties associated with using organic scintillators for the prompt detection of neutrons below an energy of $\sim$100 kev have resulted in the widespread use of "nuclear reaction" detectors. The most suitable reactions are $B^{10}(n,\alpha)Li^7$ + 480 kev γ + 2.3 Mev and $Li^6(n,\alpha)T$ + 4.8 Mev. The various detectors incorporating these reactions have been reviewed by Muehlhause in Chapter III.B.

Pulsed Van de Graaff machines have recently been developed to produce neutron bursts of a few nanoseconds duration (Go61).

The need to detect neutrons of less than 100 kev efficiently and with a time resolution of several nanoseconds is therefore apparent. The most successful prompt neutron detector in the neutron energy range under consideration has been the NaI(Tl)-B^{10} detector of Rae and Bowey (Ra53), subsequently developed by Good, Neiler, and Gibbons (Go58). This detector involves the observation of the 480-kev γ ray from the B^{10}($n,\alpha\gamma$)Li^7 reaction. Good *et al.* (Go58) obtained a time resolution of $\sim$6 nanoseconds using a "fast-slow" coincidence system; the figure of 6 nanoseconds appears to be an electronic rather than an intrinsic limitation.

The work of Voitovetskii, Tolmacheva, and Arsaev (Vo59), and Bollinger, Thomas, and Ginther (Bo59) demonstrated that cerium-activated glass scintillators loaded with either lithium or boron are suitable for detecting slow ($<$10 ev) neutrons. Recently, Firk, Slaughter, and Ginther (Fi61) have reported the use of a Li^6-loaded glass scintillator for detecting neutrons up to $\sim$100 kev energy using nanosecond time-of-flight techniques. A detector time resolution of $<$2 nanoseconds was achieved.

A. *Preparation of Glass Scintillators*

(1) **Chemical Composition.** The compositions of a large number of boron- and lithium-loaded glass scintillators have been investigated by Ginther (Gi58, Gi60), with the objective of increasing the boron and lithium content, and simultaneously increasing the pulse height and a resolution of the (n,α) peaks.

A typical composition used successfully for detecting neutrons up to 100 kev energy is as follows:

Li^{6}O$_{1/2}$	0.355
AlO$_{3/2}$	0.075
CeO$_{3/2}$	0.02
SiO$_{3/2}$	0.55

where the numbers are molar ratios. This glass has the optical properties of those given by Ginther (Gi60).

(2) **Manufacture.** Details of the manufacture of glass scintillators have been given in a number of papers (Gi58, Gi60, Bi61, Fi61). The method used by Ginther consists of heating high purity ingredients to 1500°C. in a platinum crucible. The process is carried out

66

2244

in a reducing atmosphere to ensure that the cerium is in the cerous ion form necessary for activating the scintillator.

B. Neutron Detection Properties of Glass Scintillators

(1) **Pulse-Height Spectra from the Reactions $Li^6(n,\alpha)T$ + 4.8 Mev and $B^{10}(n,\alpha)Li^7$ + 480-kev γ + 2.3 Mev.** Pulse-height spectra from lithium- and boron-loaded glasses, when irradiated with low energy neutrons (Fi61), are shown in Fig. 5. For the lithium glass, the well-defined peak with a full width at half maximum of 25 per cent is readily selected using a single-channel pulse-height analyzer. The equivalent electron energy of the $Li^6(n,\alpha)T$ peak is 1.6 Mev.

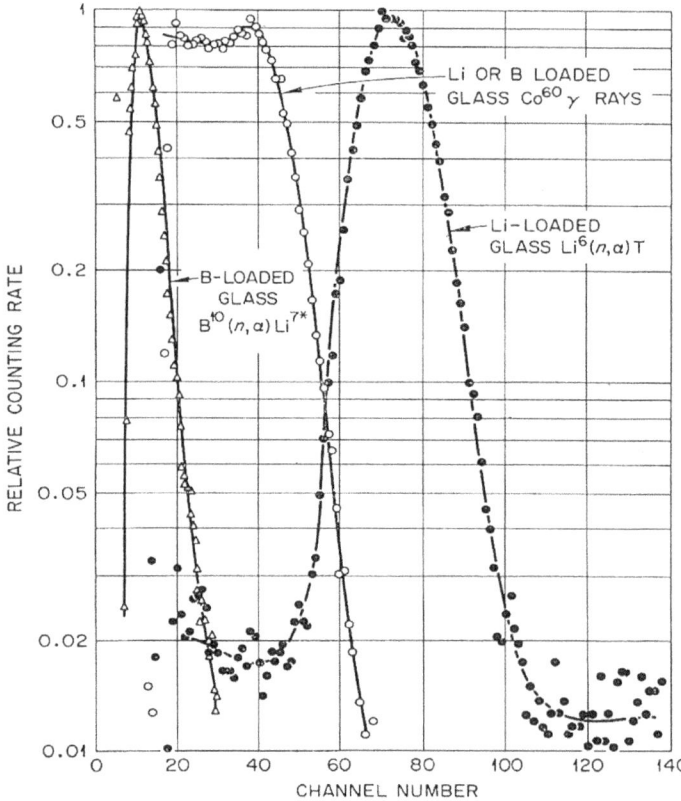

Figure 5. Typical pulse-height spectra in lithium- and boron-loaded glass scintillators (Fi61).

Firk *et al.* (Fi61) concluded that the 2.1-Mev α particle contributes 400 kev and the 2.7-Mev triton contributes the remaining 1.2-Mev equivalent electron energy. This large difference in response to α particles and tritons is atypical of crystalline inorganic scintillators. The glass, however, does not have their regular structure and is therefore likely to exhibit unusual scintillation properties.

In the boron glass the $B^{10}(n,\alpha)Li^7$ peak occurs at an equivalent electron energy of 250 kev as shown in Fig. 5. The full width of the peak at half maximum is 50 per cent. It is not so surprising, in view of the above mentioned response to α particles and tritons, that the products of the $B^{10}(n,\alpha)Li^7$ reaction, with their high specific ionization, contribute a total equivalent electron energy of only 250 kev.

(2) **Gamma-Ray Sensitivity.** Ginther and Schulman (Gi58) have shown that cerium-activated glasses are useful detectors of γ rays. This is clearly a disadvantage when used for neutron detection. However, in the case of lithium glass, the pulse height of the (n,α) peak is such that γ rays of energy <1.3 Mev make a negligible contribution. This is demonstrated in Fig. 5 where the pulse-height spectrum from Co^{60} γ rays is compared with the (n,α) peak. The mass absorption coefficient for 1.3-Mev γ rays of a lithium glass having the composition given in Section 3.A(1) is ~0.055 cm^2/gm.

The low pulse height of the (n,α) peak in boron glass can be a serious drawback in many applications, since γ rays of only ~100 kev contribute to the (n,α) peak.

(3) **Time Characteristics of the Scintillation Pulse.** An important feature of a scintillator is the decay time of the light pulse. For cerium-activated glasses (both lithium- and boron-loaded) the main component has a decay time of ~100 nanoseconds (Bo59,Fi61).

(4) **Neutron Detection Efficiency.** The Li^6-loaded glass scintillator used by Firk *et al.* (Fi61) had an efficiency of 20 per cent for detecting 1-kev neutrons with a $1/v$ dependence up to ~50 kev. So far, boron-loaded glasses have been prepared containing only natural boron. The use of B^{10} in such glasses should enable efficiencies ~50 to 60 per cent to be attained for detecting 1-kev neutrons using a scintillator 1 cm thick.

C. Applications to Nanosecond Time-of-Flight Measurements

The first use of a Li^6-loaded glass scintillator for detecting 5- to 80-kev neutrons using nanosecond time-of-flight techniques has re-

68

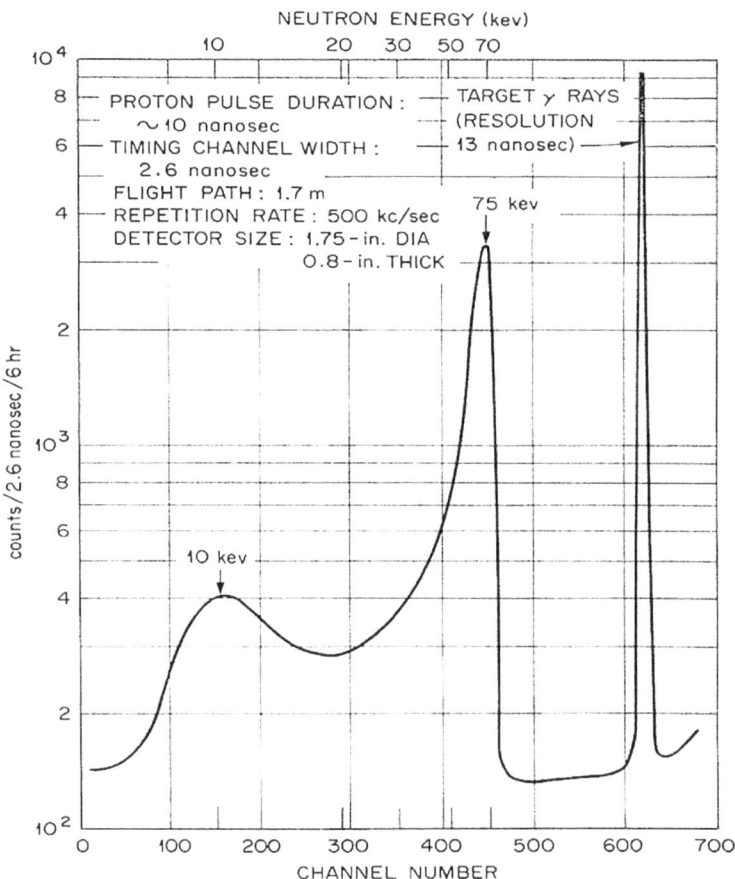

Figure 6. The observed neutron energy spectrum from the Li⁷(p,n)Be⁷ reaction using a Li⁶-loaded glass scintillator and nanosecond time-of-flight techniques. The time resolution of the glass scintillator was less than 2 nanoseconds (Fi61).

cently been reported (Fi61). These measurements were made using the Oak Ridge 3-Mv pulsed Van de Graaff machine and associated fast neutron time-of-flight spectrometer. A typical observed neutron energy spectrum from the Li⁷(p,n)Be⁷ reaction, for protons 15 kev above the reaction threshold, is shown in Fig. 6. A time resolution of <2 nanoseconds for the detector was determined by observing 1.4- to

1.8-Mev γ rays (equivalent to the $Li^6(n,\alpha)T$ peak) from the target of the Van de Graaff machine when using a beam-pulsing system which provided a burst duration of <2 nanoseconds (Go61). The efficiency of the Li^6-loaded glass scintillator was comparable with that of a 1-cm thick $B_4^{10}C$ slab and 4-in. diameter $\times$ 2-in. thick NaI(Tl) crystal previously used (Chapter IV.A). The lithium glass, however, has improved time resolution.

The low pulse height and relatively poor resolution of the $B^{10}(n,\alpha)$-Li^7 peak preclude the use of boron-loaded glasses from nanosecond time-of-flight measurements. The high neutron efficiencies obtainable with boron glasses would, however, make them suitable for many experiments not requiring excellent time resolution.

References

(Ba60)   Batchelor, Gilboy, Purnell, and Towle, *Nuclear Instr. and Methods* **8,** 146 (1960).

(Bi61)   A. M. Bishay, *J. Am. Ceramics Soc.* (in press).

(Bl60)   Bloom, Kaifer, and Schrader, *I.R.E. Trans. on Nuclear Sci.* NS-7, 170 (1960).

(Bo59)   Bollinger, Thomas, and Ginther, *Rev. Sci. Instr.* **30,** 1135 (1959).

(Bo61)   L. M. Bollinger and G. E. Thomas, *Rev. Sci. Instr.* (in press).

(Br56)   F. D. Brooks, "Organic Scintillators," in *Progress in Nuclear Physics.* Vol. 5, O. R. Frisch, ed., Pergamon Press, London, 1956, pp. 252–313.

(Br56a) F. D. Brooks, Harwell, private communication (1956).

(Br58)   F. D. Brooks, *Liquid Scintillation Counting*, C. B. Bell and F. N. Hayes, eds., Pergamon Press, London, 1958, p. 268.

(Br59)   F. D. Brooks, *Nuclear Instr. and Methods* **4,** 151 (1959).

(Br59a) F. D. Brooks and E. R. Rae, "Variation of $\eta/\bar{\nu}$ for U^{235}, *Atomic Energy Res. Establ. (Harwell) Report* NRDC 123 (1959).

(Br60)   Brooks, Pringle, and Funt, *I.R.E. Trans. on Nuclear Sci.* NS-7, 35 (1960).

(Da61)   W. Daehnick and R. Sherr, *Rev. Sci. Instr.* **32,** 666 (1961).

(Fi61)   Firk, Slaughter, and Ginther, *Nuclear Instr. and Methods* (in press)

(Fo58)   M. Forte, *Proc. Intern. Conf. Peaceful Uses Atomic Energy A/Conf.* 15/p/1514 (1958).

(Gi58)   F. J. Ginther and J. H. Schulman, *I.E.E. Trans. on Nuclear Sci.* NS-5, 92 (1958).

(Gi60)   R. J. Ginther, *I.R.E. Trans. on Nuclear Sci.* NS-7, 28 (1960).

(Go58)   Good, Neiler, and Gibbons, *Phys. Rev.* **109,** 926 (1958).

(Go61)   W. M. Good and R. F. King, private communication (1961).

(Li59)   Litherland, Almqvist, Batchelor, and Gove, *Phys. Rev. Letters* **2,** 104 (1959).

(Ne61)   J. H. Neiler and V. Pethe, ORNL, private communication (1961).

(Ow58)   R. B. Owen, *I.R.E. Trans. on Nuclear Sci.* NS-5, 198 (1958).

(Ow59)   R. B. Owen, *Nucleonics* **17,** 92 (1959).

2248   F. W. K. FIRK

(Ra53)   E. R. Rae and E. M. Bowey, *Proc. Phys. Soc.* **66A,** 1073 (1953).
(Re61)   Rethmeier, Boersma, and Jonker, *Nuclear Instr. and Methods* **10,** 240
          (1961).
(Sl61)   Slaughter, Firk, and Ginther, *Bull. Am. Phys. Soc.* **6,** 275 (1961).
(Vo59)   Voitovetskii, Tolmacheva, and Arsaev, *Atomnaya En.* **6,** 321 and 472
          (1959).
(Wr56)   G. T. Wright, *Proc. Phys. Soc.* **69B,** 358 (1956).

NEUTRON AND PHOTON REACTION STUDIES USING LOW ENERGY ELECTRON LINEAR ACCELERATORS

F. W. K. FIRK

Nuclear Physics Division, Harwell, England

The role of electron linacs in two branches of nuclear reactions is discussed. Firstly, in the determination of neutron resonance parameters ($E_n < 100$ keV) and secondly, in the measurement of energy spectra of photo-particles ($E_\gamma < 50$ MeV). Results of high resolution (γ,n) and (γ,p) experiments emphasize the "nuclear spectroscopy" aspect of photo-reactions in light nuclei.

1. Introduction

In recent years, a considerable amount of information has been obtained in the fields of low energy neutron spectroscopy and photonuclear reactions using electron linacs. At the previous accelerator conference of this series, Leon Katz discussed the desirable features of a 100 MeV electron linac and the experiments which might be carried out with it[1]. In the next forty minutes I should like to discuss in detail the work performed by low-energy electron linac groups during the last ten years.

A brief historical survey of these machines provides an interesting background to the subject. R.f. travelling-wave electron linacs were developed simultaneously and independently in England and the U.S. between 1944 and 1947[2,3]. A number of papers followed, giving the theoretical and experimental details of these machines[4-6]. At this early stage of development, the physics interest followed two quite different courses. At Stanford, the interest was clearly in very high energies at relatively low electron currents whereas at Harwell the emphasis was on low energies and high currents. The outlook in England was, no doubt, influenced by the suggestion of Cockcroft[7,8] that these low energy machines could be used for providing intense pulsed sources of neutrons via (γ,n) and (γ,f) reactions. The primary neutron spectrum is degraded by a suitable moderator in order to obtain an enhanced flux of low energy neutrons. Conventional time-of-flight methods are then used to determine the neutron energies. The various accelerators in the field of slow neutron spectroscopy are listed in table 1; the relevant performance figures are included. In particular, I should like to draw your attention to the actual running hours achieved by some of them: they represent a remarkable achievement on the part of the designers and operating staff of these complex devices. It is clear that the development of these low energy, high powered machines largely took place in Europe during the period 1950–1958 and that only during the last five years or so have they become popular in the United States.

The use of these machines for studying photo-reactions has developed more slowly than their use in slow neutron studies. The interest has, however, been more evenly spread throughout the various laboratories. An important development was made by the M.I.T. group in the middle 1950's (using a uniform standing-wave accelerator) in which they measured the energy spectra of photoneutrons using a nanosecond time-of-flight technique[19]. At Harwell, we have further developed this technique: the energy range covered is $15 \lesssim E_\gamma \lesssim 35$ MeV[20]. This method is also being used now at R.P.I.[21] with incident photon energies up to 60 MeV and in Japan[22] with values of $E_\gamma < 25$ MeV.

The production of nearly monochromatic γ-ray beams has been achieved at Saclay[23], Livermore[24] and San Diego[25] and impressive (γ,n) cross sections have recently been obtained in the energy range below 30 MeV: typical resolutions are 300–500 keV.

The successful use of solid state counters to measure energy spectra of charged particles in (γ,p) and (γ,d) reactions has recently been reported from the new Yale electron accelerator group[26].

The principle accelerators now performing photonuclear studies are listed in table 1. (This excludes the important (e, e′) reactions carried out at Stanford, Orsay and Darmstadt which are reviewed in the next paper.)

1) L. Katz, Nucl. Instr. and Meth. **11** (1961) 14.
2) Fry, Harvie, Mullett and Walkinshaw, Nature **160** (1947) 351.
3) Ginzton, Hansen and Kennedy, Rev. Sci. Instr. **19** (1948) 89.
4) W. Walkinshaw, Proc. Phys. Soc. (Lond) **61** (1948) 246.
5) R. B. R. Harvie, Proc. Phys. Soc. (Lond) **61** (1948) 255.
6) Mullett and Loach, Proc. Phys. Soc. (Lond) **61** (1948) 271.
7) J. D. Cockcroft, Nature **163** (1949) 869.
8) Duckworth and Merrison, Nature **163** (1949) 869.
19) Bertozzi, Demos and Sargent, private communication.
20) Firk, Whittaker, Bowey, Lokan and Rae, Nucl. Instr. and Meth. **23** (1963) 141.
21) P. F. Yergin, private communication (1963).
22) M. Kimura, private communication (1963).
23) Schuhl and Tzara, Nucl. Instr. and Meth. **10** (1961) 217.
24) Jupiter, Hansen, Shafer and Fultz, Phys. Rev. **121** (1961) 866.
25) Sund and Walton, private communication (1963).
26) O'Connell and Bockelman, private communication (1963).

F. W. K. FIRK

TABLE 1

Electron linacs used for slow neutron spectroscopy and photonuclear studies

Location	Operational date for experiments	Energy (MeV)	Pulse current (mA)	Pulse duration (μsec)	Neutrons per sec in the pulse	Operating hours/week	References
Harwell, Eng.	1949–52	3.5	120	2	$\sim 10^{12}$		8, 9
Harwell, Eng.	1952–59	15	25	0.2–0.8	$\sim 10^{14}$	120	10, 11, 13
Saclay, France	1958–	26	400	0.07–1	$\sim 10^{16}$	~ 80	14
	(1963)	(30)	(500)	(0.01)			
Moscow, U.S.S.R.	1958–	30	400	0.05–0.6	$\sim 10^{16}$		15
Harwell, Eng.	1959–	35†	600	0.1–1.5	$\sim 10^{17*}$	~ 110	13
Harwell, Eng.	(1961)	(40)	(~ 1000)	(0.005)			20, 95
Livermore, U.S.A.	1959	30	400	0.1–1.5	$\sim 10^{16}$		16
San Diego, U.S.A.	1959–61	20	200	0.5–2			17
Tokyo, Japan	1959–	20	150	1			
Tokyo, Japan	(1963)	(25)	(200)	(0.04)			22
San Diego, U.S.A.	1961–	28	700	0.1–5	$\sim 10^{16}$	> 100	17
San Diego, U.S.A.	(1963)	(30)	(~ 1000)	(0.02)			17
R.P.I., U.S.A.	1961–	60	> 1500	0.1–6	$> 10^{17}$	60	18
R.P.I., U.S.A.	(1963)	(> 65)	(> 1500)	(0.01)			21
Yale, U.S.A.	1962	40	> 1000	0.1–6	$\sim 10^{17}$		18
M.I.T.§, U.S.A.	~ 1954	18	100	0.004			19
Yale§, U.S.A.	~ 1951–60	10	100	1.2	$\sim 10^{14}$		12

* Includes a factor of 10 from sub-critical U^{235} target.

§ These machines are standing-wave accelerators. The fast pulse performances are given in parentheses.

† *Note added in proof*: now operating at 45 MeV.

2. Neutron interactions in the resonance region

$(E_{neut} \lesssim 100 \text{ keV})$

The fact that slow neutron resonance parameters provide some of the most accurate and unambiguous information available on the properties of nuclear energy levels is well known. The energy region of interest is limited to that just above the neutron binding energy. Nevertheless, the range is sufficiently wide to obtain many details of the systematics of energy levels; for example: the distributions of reduced neutron widths, fission widths, total radiation widths and partial radiation widths. Other important quantities which emerge are the s- and p-wave strength functions and the distribution of level spacings. On reading the reviews of these topics given at Conferences in recent years[27,28] one might conclude that the subject is well understood. For instance, it is now known that the distribution of reduced neutron widths is close to a Porter-Thomas form ($\nu = 1$), the distribution of partial radiation widths probably has $\nu = 1$, the distribution of fission widths has a small value of ν, the total radiation widths are reasonably constant from resonance to resonance in a single nuclide and the s-wave strength function exhibits large fluctuations at $A \approx 52$ and $A \approx 155$. There are, however, many additional questions which we might ask:

1. Does the distribution of reduced neutron widths ever depart from a Porter-Thomas form?

2. Do the fission width distributions fall into two groups depending upon the resonance spins?

3. Are the total radiation widths spin dependent and do they vary systematically with the angular momentum of the captured neutron?

4. Does the distribution of partial radiation widths always have $\nu = 1$?

5. Is the s-wave strength function spin-dependent?

6. Does the p-wave strength function split at $A \sim 100$?

7. What is the value of the spin cut-off factor σ^2 in the expression for level density?

9) Merrison and Wiblin, Proc. Royal Soc. **A215** (1952) 278.
10) Bareford and Kelliher, Philips Tech. Rev. **15** (1953) 1.
11) Firk, Reid and Gallagher, Nucl. Instr. **3** (1958) 309.
12) Schultz and Wadey, Rev. Sci. Instr. **22** (1951) 383.
13) Poole and Wiblin, Proc. Inter. Conf. Geneva, **14** (1958) 266.
14) Leboutet, Picard and Vastel, L'Onde Electrique **37** (1957) 28.
15) Vladimirsky and Sokolovsky, Proc. Inter. Conf. Geneva,**14** (1958) 283.
16) Austin and Fultz, Rev. Sci. Instr. **30** (1959) 284.
17) J. R. Beyster, private communication (1961).
18) Nygard and Post, Nucl. Instr. and Meth. **11** (1961) 126.
27) D. J. Hughes, Nuclear Reactions Vol. 1 (North-Holl. Publ. Comp., Amsterdam, 1956).
28) J. A. Harvey, Proc. Kingston Conf. (North-Holl. Publ. Comp./ Univ. of Toronto, 1960).

73

8. Does the distribution of level spacings always fit the Wigner-Mehta theory?

9. Do "direct-capture" processes play an important part in the resonance region?

10. To what extent is the potential scattering length spin-dependent? etc. etc.

From the experimental stand-point, the answers to all these questions are, as yet, unknown.

In the golden era of fast neutron choppers (roughly 1952 to 1958) three important facts were reasonably well established, namely: the Porter-Thomas form of the reduced neutron width distribution, the variation of the s-wave neutron strength function for $50 \lesssim A \lesssim 240$ and its connection with the optical model and the distribution of level spacings.

Since 1958 the chopper groups have been finding it increasingly difficult to maintain their pre-eminence in this field. This is due to the considerable experimental difficulties associated with answering any one of the questions outlined above. The need for very high resolution coupled with high neutron intensities has resulted in the popularity of bigger and better electron linacs.

We now turn to a discussion of the part played by these machines in helping to answer some of the above questions. Before describing recent experiments, it will be useful to remind you of the basic parameters which arise in neutron resonance measurements. They are the resonance energies E_λ, the total widths Γ, the partial widths Γ_γ, Γ_n and Γ_f, the spins J, the potential scattering lengths a and the level spacings. The aim of experiments is to determine these quantities and, where appropriate, obtain their mean values and dispersions about their means. These statistical properties are then compared with predictions made using various nuclear models.

In many experiments in this field, the result is obscured by two factors: the Doppler broadening due to thermal motion of the target nuclei and the resolution broadening due to inadequate instrumental resolving power. Both these factors create difficulties in extracting the true resonance parameters; however, methods are available for correcting for these effects[29,30] and they will not be pursued here.

The area under an isolated cross section curve is independent of the resolution effect and is, therefore, a widely used method of analysis. By making suitable corrections for Doppler broadening, the following results are obtained:

a) from total cross sections

$$A_\mathrm{T} = 2\pi^2 \lambdabar^2 g \Gamma_\mathrm{n} \qquad (1)$$

b) from capture cross sections

$$A_\gamma = 2\pi^2 \lambdabar^2 g \Gamma_\mathrm{n} \Gamma_\gamma / \Gamma \qquad (2)$$

and c) from scattering cross sections

$$A_\mathrm{n} = 2\pi^2 \lambdabar^2 g \Gamma_\mathrm{n}^2 / \Gamma \qquad (3)$$

Measurements of the total, scattering and capture cross sections, therefore, give values of $\Gamma_\mathrm{n}, \Gamma_\gamma$ and the statistical weight factor $g(=[2J+1]/[2(2I+1)])$. Here, I is the ground state spin of the target nucleus and J is the spin of the resonance.

Although numerous values of $g\Gamma_\mathrm{n}$ have been obtained from total cross section measurements very few complete analyses have been obtained by measuring all three quantities[31-35].

Apart from important factors of neutron intensity and resolving power, electron linacs are well-suited to

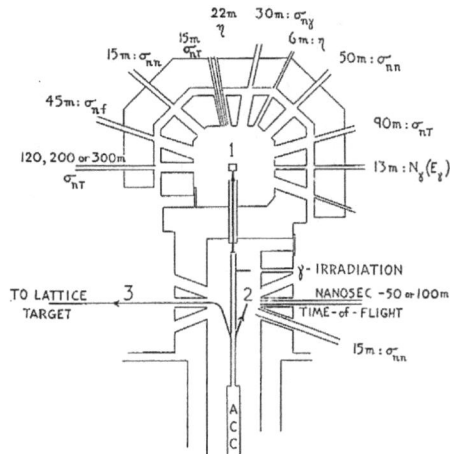

Fig. 1. The layout of the neutron time-of-flight experiments now in operation on the Harwell 35 MeV electron linac.

the simultaneous measurement of the different experiments mentioned above. Fig. 1 shows an example of the wide variety of work now being carried out on the Harwell linac. There are up to ten experiments running simultaneously, each with their associated 4096 channel magnetic tape time-of-flight analysers. Flight path

29) D. J. Hughes, Journ. Nucl. Energy 1 (1957) 237.
30) Lynn and Rae, Journ. Nucl. Energy 4 (1957) 418.
31) Rae, Collins, Kinsey, Lynn and Wiblin, Nuclear Phys. 5 (1957) 89.
32) Evans, Kinsey, Waters and Williams, Nuclear Phys. 9 (1958) 205.
33) F. W. K. Firk, Nuclear Phys. 9 (1958) 198.
34) Firk and Moxon, Nuclear Phys. 12 (1959) 552.
35) Waters, Evans, Kinsey and Williams, Nuclear Phys. 12 (1959) 563.

lengths vary from 6 to ~ 300 metres. In addition to the main facility of the U^{235} booster, the electron beam may be deflected into two other positions either for (γ, n) measurements or into a new target cell in which Michael Poole and collaborators[36]) carry out measurements of energy spectra of neutrons from lattices which simulate reactor arrangements. The main target experiments and the lattice experiments operate at the same time by pulsing the electron beam into either target position at any desired repetition rate and pulse length.

2.1. TOTAL CROSS SECTION MEASUREMENTS

At the Kingston Conference in 1960[28]) our results were used to illustrate high resolution measurements on U^{238} in the energy region up to 2 keV. About 100 resonances were analysed and provided a useful test of the Porter-Thomas hypothesis for the distribution of reduced neutron widths and of the Wigner-Mehta form of the distribution of level spacings[37,38]). Fig. 2 shows the results. It is seen that the Porter-Thomas distribution is a good fit to the data and the Wigner-Mehta distribution looks reasonably sound. The problems of testing the Porter-Thomas distribution more accurately are clearly severe. The difficulties are mainly concerned

with resolving and analysing the small levels and of covering a sufficiently wide energy range in order that one can be certain that no serious deviations are likely to occur for values of $\Gamma_n^0/\bar{\Gamma}_n^0 \gtrsim 10$, say.

Likewise, the level spacing distribution agrees reasonably well with that of Wigner and Mehta but, with such a small sample of 100 resonances, it is by no means a

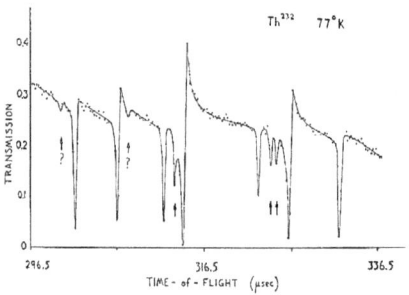

Fig. 3. The observed transmission of neutrons through a Th^{232} sample cooled to 77°K [40]) (resolution 1.7 nsec/m, energy ~ 500 to 600 eV).

settled problem. Recent results from the Columbia synchro-cyclotron indicate some anomalies in the Th^{232} distribution[39]). In this connection, an interesting result has recently been obtained at Saclay[40]) in which the total cross section of Th^{232} has been measured with high resolution using a thick sample cooled to 77°K (fig. 3). In the first keV they observe ~ 30% more levels compared with uncooled experiments of comparable resolution[41]). This is a clear warning to those of us who have produced level spacing distributions in the past. Let us hope the additional levels are all p-wave!

In a few cases, where the resolution is adequate, a direct shape analysis of a cross section curve is possible[42-44]). The total neutron cross section of vanadium is such a case and we have measured it in the energy region 600 eV to 25 keV (see fig. 4). The resolution (3 nsec/m) and accuracy of the absolute cross section

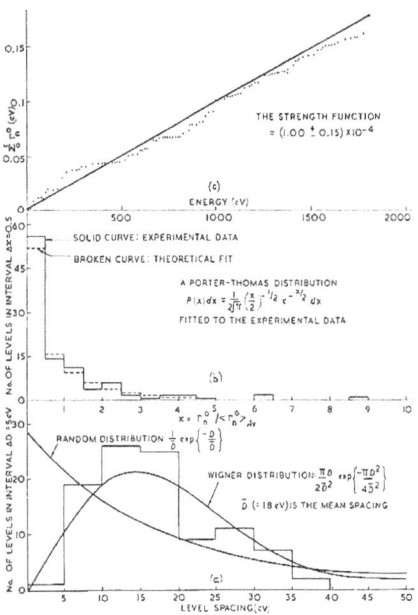

Fig. 2. Results obtained for 100 resonances in U^{238} + n [28]). (a) The distribution of level spacings. (b) The distribution of reduced neutron widths. (c) The neutron strength function.

36) Poole, Schofield and Sinclair, Proc. I.A.E.A. Conf. Amsterdam (1963).
37) E. P. Wigner, Proc. Gatlinburg Conf. on Neutron Time-of-Flight (1956).
38) M. L. Mehta, Nuclear Phys. 18 (1960) 395.
39) W. W. Havens, private communication (1963).
40) Michaudon and Ribon, private communication (1963).
41) Uttley and Jones in Neutron Time-of-Flight Methods (E.A.N.D.C. Brussels 1961).
42) Bollinger, Dahlberg, Palmer and Thomas, Phys. Rev. 100 (1955) 126.
43) Lynn, Firk and Moxon, Nuclear Phys. 5 (1958) 603.
44) Firk, Lynn and Moxon, Proc. Phys. Soc. (London) 82 (1963).

are sufficient to permit a shape analysis of the entire cross section using a general multi-level expression based on the Thomas approximation[45] of the Wigner R-matrix theory[46]). The spins of the levels are either $J = 3$ or $J = 4$ and are directly determined from the peak cross sections for the largest levels. The most

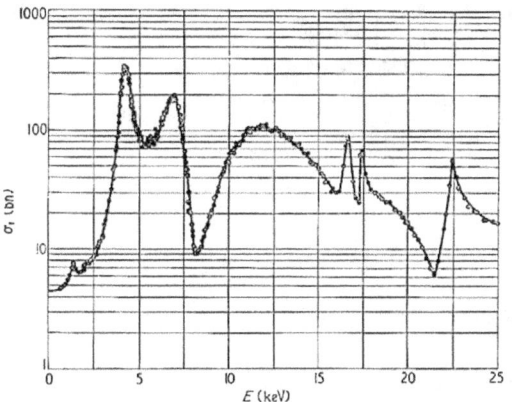

Fig. 4. The observed total neutron cross section of natural vanadium up to 25 keV. Apart from the region of the two narrow resonances at about 17 keV, the cross section is the true cross section uneffected by Doppler or resolution broadening[44].

significant part of the analysis is the determination of the values of R_J^∞, i.e. the contribution of distant levels for particular spins J. It is then possible to determine the potential scattering lengths

$$a_J = a(1 - R_J^\infty) \quad (\text{see Feshbach et al.}[47]))$$

(a is the "nuclear radius" and is proportional to $A^{\frac{1}{3}}$). Furthermore, the magnitude and sign of R_J^∞ give a good

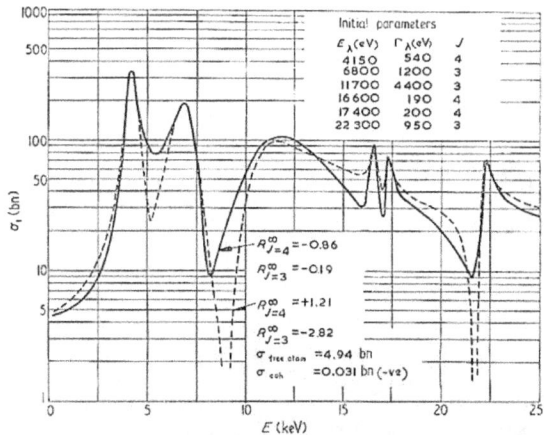

Fig. 5. Initial multi-level fit to the vanadium cross section showing the sensitivity to the values of R_J^∞/J [44].

indication of the location of the 3s single particle state in V [52]). Fig. 5 shows the initial multi-level fit to the data using the two possible pairs of values of R_J^∞ (consistent with the thermal scattering lengths). Clearly, the pair with negative values give the better fit. In a final, least-squares analysis, a slight energy dependence is included to allow for nearby levels in the range 25–100 keV. The final result indicates that the scattering lengths a_J differ for the two spin states, namely,

$$a_{J=3} = 7.5 \text{ fm}$$
$$a_{J=4} = \gtrsim 5 \text{ fm}$$

Also, it appears that the 3s giant resonance is located close to the neutron binding energy. In early optical model calculations the 3s state was set at zero neutron energy at $A = 55$.

By similar careful analyses of all nuclei in the region $A = 50$ to 60 it should be possible to obtain important information in connection with optical model calculations.

2.2. CAPTURE CROSS SECTION MEASUREMENTS

Until recently, very few capture cross section measurements had been made in the resonance energy region. Early measurements using the late 15 MeV machine at Harwell were limited to energies below about 200 eV and to the determination of radiation widths[48]). The capture cross section between resonances and the average capture cross section in the keV region were not measurable above the background. Since 1960, this position has changed considerably. Measurements at Harwell[49]), R.P.I.[50]), San Diego[51]), and Saclay[52]) have produced greatly improved data in the energy range from a few eV to about 100 keV. This change is due to two main reasons: the resolving power and neutron intensity of these accelerators coupled with the development of improved γ-ray detectors. Moxon[49]) is using a detector in which electrons produced in a thick graphite γ-ray to electron converter are observed with a plastic scintillator. The efficiency of this device is essentially independent of the multiplicity of capture γ-rays. Furthermore, the efficiency for detecting γ-rays ε_γ, compared with the efficiency for detecting neutrons is very favourable and is given by $\varepsilon_\gamma/\varepsilon_n > 6000/1$.

At R.P.I.[50]), San Diego[51]) and Saclay[52]) large liquid

45) R. G. Thomas, Phys. Rev. 97 (1955) 224.
46) Wigner and Eisenbud, Phys. Rev. 72 (1947) 29.
47) Feshbach, Porter and Weisskopf, Phys. Rev. 96 (1954) 448.
48) Rae and Bowey, Journ. Nucl. Energy 4 (1957) 179.
49) Moxon and Rae, Nucl. Instr. and Meth. 24 (1963) 445.
50) Block, Russel and Hockenbury, Bull. A.P.S.
51) E. Haddad, private communication (1963).
52) J. Julien, private communication (1963).

VI. EXPERIMENTAL TECHNIQUES

76

F. W. K. FIRK

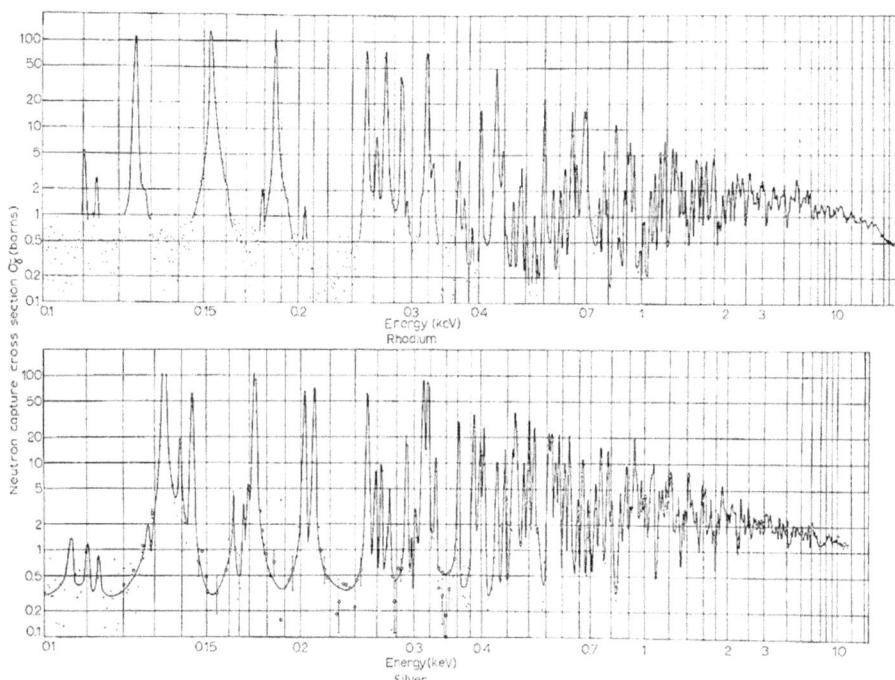

Fig. 6. The observed capture cross sections of Ag and Rh in the region ~ 1 eV to ~ 50 keV [53].

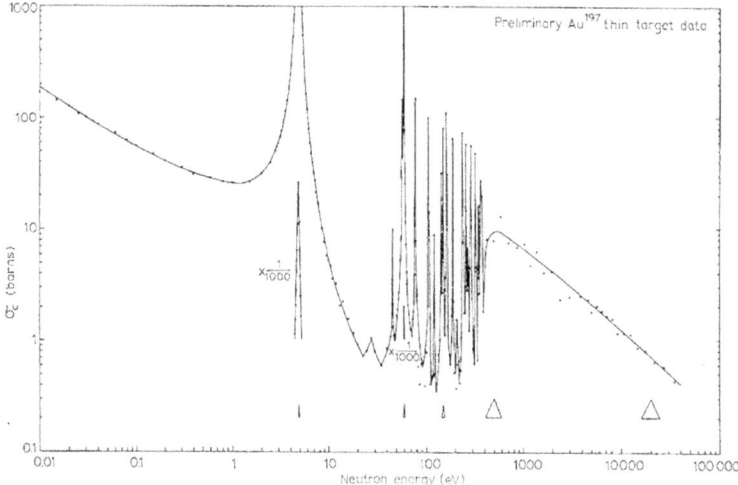

Fig. 7. The observed capture cross section of Au obtained with a large (4000 litre) liquid scintillator [51]. The energy range of a single measurement is from 0.01 eV. (This is a preliminary result and the absolute cross section should not be quoted without consulting Haddad [51]).

scintillators are used to detect the γ-rays. These devices have high efficiencies (and high backgrounds!) – and in the case of the San Diego detector (4000 litres) a good pulse height spectrum for 8 MeV γ-rays. The data from these experiments provide values of radiation widths with an accuracy previously unattainable in this field. Table 2 lists examples for radiation widths in $U^{238} + n$.

TABLE 2

Resonance Energy (eV)	Γ_γ (meV)	Γ_n (meV)	$\sigma_0\Gamma^2$ (b.eV²)
36.53	23.2 ± 3.5	34.5 ± 3.0	138.6 ± 6
65.70	24.1 ± 2.0	25.5 ± 1.5	49.6 ± 3.2
80.35	—	1.8 ± 0.3	—
101.9	24.1 ± 2.0	69 ± 3	162 ± 3.2
116.2	22.1 ± 1.3	37.4 ± 3.5	50.5 ± 4.2
144.8	—	0.68 ± 0.07	—
164.4	—	2.80 ± 0.25	—
188.8	22.7 ± 2.2	152 ± 10	368 ± 12
207.5	23.8 ± 2.8	57 ± 5	56 ± 9
236.0	24.8 ± 3.2	40 ± 6	—
271.9	(23.2)	22 ± 7	—
289.0	(23.2)	16 ± 5	—
346.4	26.5 ± 3.5	44 ± 5	23 ± 3.5
374.7	—	1.2 ± 0.2	—
408.5	(23.2)	21.5 ± 5	—
432.5	(23.2)	16 ± 3	—
446.1	—	1.4 ± 0.4	—
460.4	—	6.3 ± 0.9	—
475.5	—	3.8 ± 0.8	—
515.4	27.0 ± 4.0	.49 ± 10	18.9 ± 4
532.0	20.0 ± 3.5	98 ± 40	59 ± 20
576	27.6 ± 5	49 ± 10	17 ± 3.5
589	25.0 ± 2.5	89 ± 17	45 ± 9
617	(23.2)	37 ± 10	—
625	—	2.8 ± 0.5	—
658	22.2 ± 2.5	192 ± 40	162 ± 30

obtained by Moxon[53]) in which the capture data are combined with the transmission data. Using such devices we may look forward in the immediate future to a wealth of data on the systematics of total radiation widths.

Another important point regarding these results is the ability to carry out measurements from thermal to ~ 100 keV neutron energy in a single experiment (see fig. 6). This bridges the gap between the former resonance work and the Van-de-Graaff experiments starting above 5 keV[54]). Difficulties associated with systematic errors which normally occur in the two separate energy regions are therefore eliminated. A typical example of the high quality work of Haddad[51] at San Diego is shown in fig. 7.

At the higher energies, the average capture cross sections in the range 5 to 100 keV (when combined with

transmission data) help solve the problems of the p-wave neutron strength function which has proved to be a particularly difficulty parameter to measure.

2.3. SCATTERING CROSS SECTION MEASUREMENTS

Precise measurements of the yields of elastically scattered neutrons in the resonance region are practically non-existent. A limited number of nuclei have been studied using the Harwell 15 and 30 MeV electron linacs[31,32,35,55–57]). In all these cases, the detector consisted of an annulus of BF_3 counters surrounding the target. This detector had low efficiency and poor time resolution (~ 1 μ sec.).

An improvement in this situation has now taken place at Harwell where Asghar and Brooks[58]) have successfully used Li^6-loaded glass scintillators[59]) to detect scattered neutrons with high efficiency and fast time resolution. The γ-ray background is measured simultaneously using Li^7-loaded glass scintillators. These are identical with the Li^6 glasses except that the neutron efficiency is reduced by a factor of about 100. Examples

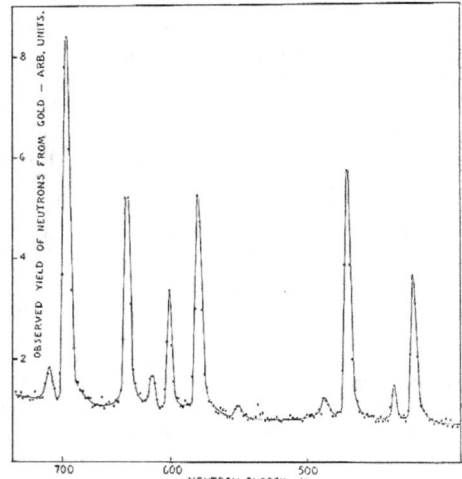

Fig. 8. The observed yield of scattered neutrons obtained using Li^6-loaded glass scintillators. The background is obtained using a Li^7-loaded glass.[58])

[53]) M. C. Moxon, private communication (1963).
[54]) Gibbons, Macklin, Miller and Neiler, Phys. Rev. 122 (1961) 182.
[55]) J. R. Waters, Phys. Rev. 120 (1960) 2090.
[56]) Frazer and Schwartz, Nuclear Phys. 30 (1962) 269.
[57]) F. Corvi, private communication (1962).
[58]) Asghar and Brooks, private communication (1963).
[59]) Firk, Slaughter and Ginther, Nucl. Instr. and Meth. 13 (1961) 313.

78

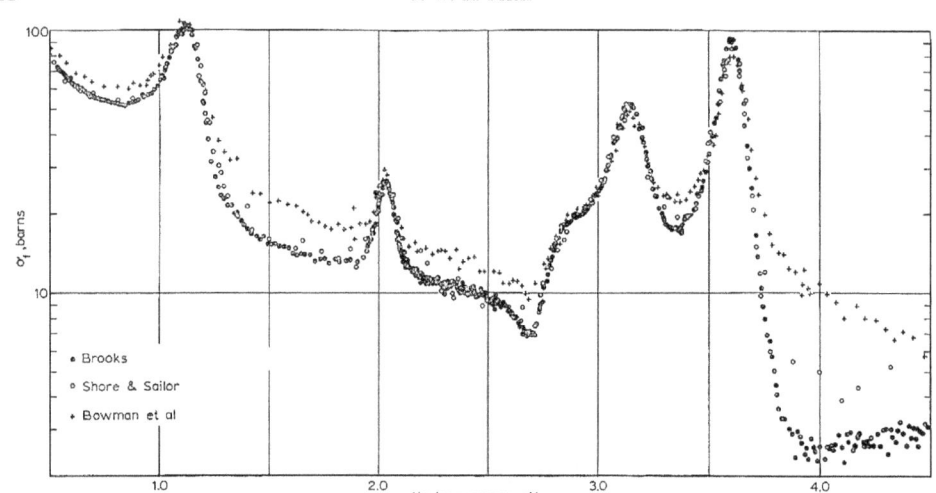

Fig. 9. The observed fission cross section of U235 obtained using a fast neutron detector (NE 213 liquid scintillator) with pulse shape discrimination to reduce the γ-ray background [63].

are shown in fig. 8 which presents the yield of neutrons from a thin gold sample. The resolution is 10 nsec/m which is an exceptional improvement compared with previous measurements.

On combining these measurements with transmission data a number of spins have been obtained: in the thulium case about 20 spin values have been established.

Brooks[60] has recently obtained results using a "bright-line scattering" technique in which the scatterer is placed close to the pulsed source of neutrons. The time-of-flight of the elastically scattered neutrons is then measured. This method has the great advantage that the capture γ-rays are separated in time from the scattered neutrons. In the case of fissile materials, in which an abundance of fission γ-rays is also produced, this technique offers the only practicable solution for measuring the scattering cross sections. Results for U235 appear very promising.

With these improved techniques for measuring spins of resonances it now seems likely that the value of σ^2 in the expression for the level density ρ_J namely

$$\rho_J = \rho_0(2J + 1) \exp \{ - (J + \tfrac{1}{2})^2/2\sigma^2 \}$$

will at last become well established by studying many resonances in which $J \gtrsim 3$ (ref. [61]).

Information on the spin-dependence of resonance parameters in general, is badly needed.

2.4. Fission cross section measurements

There are many experimental difficulties involved in measurements of fission cross sections and related parameters α, η and ν. These difficulties may be due to a number of reasons such as small available quantities of samples, intense radioactivity of samples, presence of fission and capture γ-rays, elastically scattered neutrons etc. This particular field is therefore known for its many clever techniques devised in order to minimise the background effects.

Early high resolution measurements of σ_f for U235 were made by Michaudon et al.[62] at Saclay. They used a gas scintillation counter and examined the energy range from a few eV to several hundred eV. Since then, extensive measurements of η and σ_f for U233, U235 and Pu239 have been made by Brooks[63]. The fast fission neutrons are detected in liquid scintillators using pulse shape discrimination to reduce the γ-ray background. This method of particle discrimination was developed by Brooks himself in 1956 while working on the 15 MeV electron linac[64, 65]. Fig. 9 shows the very accurate data which is now being obtained using this method. Measurements of σ_f for U235 have also been made at

[60] F. D. Brooks, private communication (1963).
[61] Firk, Lynn and Moxon, Nuclear Phys. 44 (1963) 431.
[62] Michaudon, Genin, Joly and Vendryes, Saclay Report C.E.A. 1093 (1958).
[63] F. D. Brooks, in Neutron Time-of-Flight Methods (E.A.N.D.C., Brussels 1961).
[64] F. D. Brooks, in Progress in Nuclear Physics (Pergamon Press, London, 1956).
[65] F. D. Brooks, Nucl. Instr. and Meth. 4 (1959) 151.

Livermore[66]); their absolute cross section, however, disagrees with the work of Shore and Sailor[67]) and of Brooks[68]).

Interesting experiments have recently been carried out at Harwell using gas scintillation counters for materials such as Th^{229} (Bollinger[69])) and U^{232} (James[70])). The difficulties of such measurements become clear when one considers the very small amount of sample material

made by Landon and Rae[72]) using the Harwell 15 MeV electron linac. The result was verified and the technique extended in the same year using a 2-dimensional magnetic tape recording system in which neutron time-of-flight and associated spectra of γ-rays were recorded simultaneously[73, 74]).

The significance of such experiments in obtaining fundamental information on the positions of energy levels in nuclei and the radiative transition strengths between them is borne out by the large number of groups now carrying out such experiments. These include the Harwell[75, 76]), Saclay[77]) and Yale[78]) electron linacs (as well as the chopper groups at Argonne[79]), Brookhaven[80]) and Oak Ridge[81])).

Typical results are shown in fig. 11 which shows some data on γ-ray spectra from the 98 eV Pt resonance obtained using a 2-dimensional digital analyser[82]).

The use of large NaI(Tl) crystals surrounded by anti-coincidence annuli (also NaI) at Saclay and Yale result in these improved γ-ray spectra. An interesting example is shown in fig. 12 [78]) in which the spectra from thermal

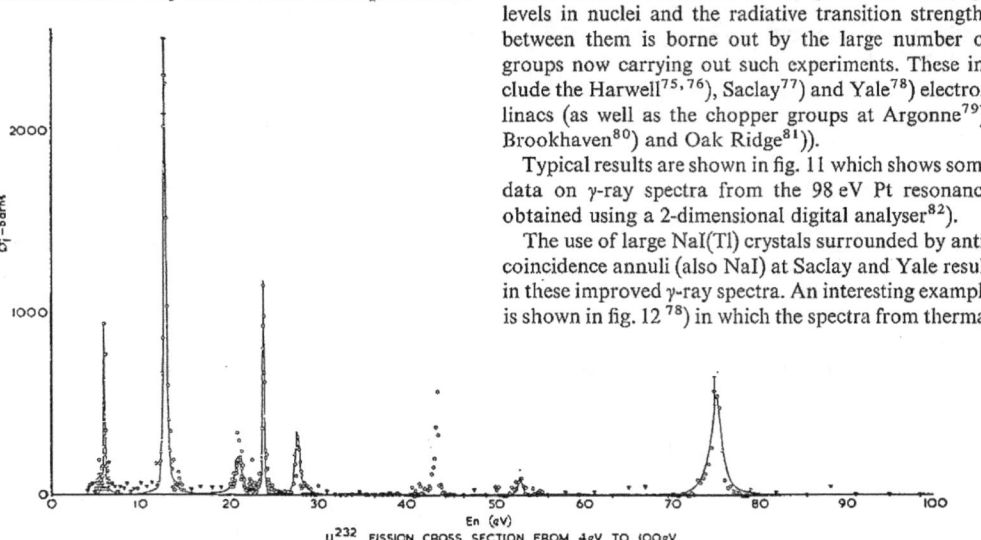

Fig. 10. The observed yield of fission fragments from U^{232} obtained using a gas scintillation counter[70]).

available (0.6 mg of Th^{229}) and the very high α-particle background from this ($> 2 \times 10^9$ α-particles per sec!).

The measurements on U^{232} [70]) are shown in fig. 10. Note the wide level spacing observed (since it is even-even nucleus) so that interference between levels is small. We may, therefore, hope to see a unique fit to this fission cross section and thereby obtain interesting data on the fission width distribution.

James[70]) has also measured σ_f for Pu^{241} from 0.009 eV to 3 keV using silicon-gold surface barrier detectors for observing the fission fragments. These measurements extend the range of energy covered compared with earlier measurements using the M.T.R. chopper[71]).

The combination of these high resolution results with σ_T and σ_n measurements will add considerably to our knowledge of the fission process.

2.5. CAPTURE γ-RAY SPECTRA

The first measurement of neutron capture γ-ray spectra in order to infer the spin of a resonance was

and resonant capture in Cu are compared. Note the very large fluctuation in the ($Cu^{63} + n$) ground state γ-ray.

66) Bowman, Auchampaugh and Fultz, Phys. Rev. 130 (1963) 1428.
67) Shore and Sailor, Phys. Rev. 112 (1958) 191.
68) F. D. Brooks, private communication (1963).
69) L. M. Bollinger, private communication (1963).
70) G. D. James, private communication (1963).
71) Simpson and Moore, Phys. Rev. 123 (1961) 559.
72) Landon and Rae, Phys. Rev. 107 (1957) 133.
73) Rae and Firk, Nucl. Instr. 1 (1957) 227.
74) Cavanagh and Boyce, Rev. Sci. Instr. 27 (1956) 1028.
75) Bird, Moxon and Firk, Nuc. Phys. 13 (1959) 525.
76) Bird and Waters, Nuc. Phys. 14 (1959) 212.
77) Julien, Corge, Huyn, Morgenstern and Netter, Physics Letters 3 (1962) 69.
78) Wasson and Draper, Physics Letters 6 (1963) 350 and private communication.
79) Carpenter and Bollinger, Nuclear Phys. 21 (1960) 66.
80) Chrien, Bolotin and Palevsky, Phys. Rev. 127 (1962) 1680.
81) Slaughter, Bird, Harvey and Chapman, Bull. A.P.S. 8 (1963).
82) F. Netter, private communication (1963).

80

F. W. K. FIRK

The distribution of partial radiation widths $\Gamma_{\gamma_i}/\bar{\Gamma}_{\gamma_i}$ has proved difficult to measure. Nevertheless, a combination of electron linac work and the precise measurements of the chopper groups[83]) now gives a value of $\nu = 1.4 \pm 0.4$ in reasonable agreement with a Porter-Thomas distribution. However, the number of transitions studied is still small and more interesting results (and anomalies) may be expected in the near future.

3. Photon induced reactions in the energy region $E_\gamma \lesssim 50$ MeV

At energies below about 30 MeV the interaction of γ-rays with nuclei results in the familiar giant electric dipole resonance the gross features of which are well known[84,85]). In medium and heavy nuclei these features include the mean energy and width of the resonance and their variation with mass number, the general shape of the (γ, n) and (γ, p) cross sections and, in some cases, the angular distributions of the emitted particles. There are so many overlapping resonances excited in these heavy nuclei that the data only provide information of a statistical nature.

In light nuclei ($A < 40$, say) the situation is, in many ways, more interesting. There are no clear-cut systematic effects regarding the position and width of the giant resonance[86]). The density of levels excited by γ-rays at energies below 30 MeV is often low enough that the energies and widths of individual levels can, in principle, be determined.

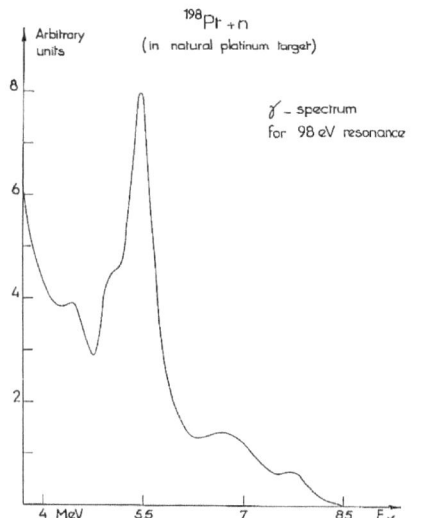

Fig. 11. The measured spectra of γ-rays from the 98 eV resonance in Pt [82]).

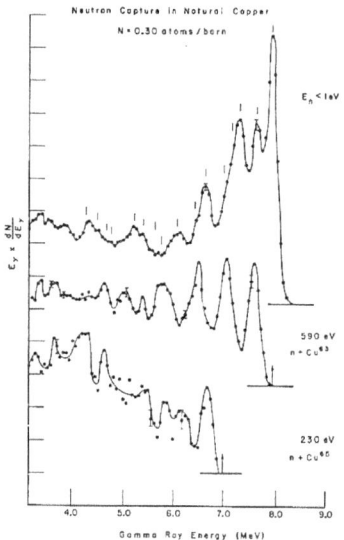

Fig. 12. The observed spectra of γ-rays from thermal and resonant capture in copper. The thermal spectrum is predominantly due to $(Cu^{63} + n)$ and may be compared with the resonance at 590 eV. Note the absence of the ground state transition in the resonance spectrum[78]).

Experimental techniques have recently been developed which allow non-overlapping levels to be resolved in many cases. The first important technical advance was made by carrying out inverse (p, γ) measurements[87-91]) with resolutions of about 20 keV (limited in general by the target thickness). In a number of cases it was also possible to select the ground state γ-ray transitions thereby obtaining the (γ, p_0) cross section using the detailed balance theorem.

During the past two years or so measurements of the energy spectra of photoparticles from (γ, n) and (γ, p) reactions have been reported using high intensity beams

Bollinger, Carpenter, Coté and Marion, Phys. Rev. (in press).
J. S. Levinger, *Nuclear Photo-disintegration* (Oxford Univ. Press, London 1960).
Fuller and Hayward, *Nuclear Reactions* Vol. II (North-Holl. Publ. Comp., Amsterdam, 1963).
E. Hayward, Rev. Mod. Phys. **35** (1963) 324.
Gove, Litherland and Batchelor, Phys. Rev. Letters **3** (1959) 177.
Tanner, Thomas and Earle, Proc. Rutherford Conf. (Heywood, London 1961).
Gemmell and Jones, private communication (1961).
Cohen, Fisher and Warburton, Phys. Rev. Letters **3** (1959) 433.
Kimura, Shoda, Mutsuro, Tohei, Sato, Kuroda, Kariyama and Akiba, Nuclear Phys. **23** (1961) 338.

from electron linacs. The excitation energies range from 15 to 60 MeV and the resolution in the (γ, n) measurements frequently exceeds that of the best charged particle experiments obtained using tandem Van-de-Graaff machines. A considerable change in the whole outlook of photoreactions is brought about by these results. The subject is no longer only a matter of measuring a smooth giant resonance extending over many MeV and in which the primary interest is in the integrated cross section. It is now a matter of classical "nuclear spectroscopy" where one is interested in the properties of individual nuclear states in much the same way as the slow neutron work just outlined or the charged particle work of tandems with which you are all familiar.

In addition to the measurements of the energy spectra of photoparticles using either fast time-of-flight (for neutrons) or magnetic spectrometers or solid state counters (for protons etc.) elegant experiments have been carried out at Saclay, Livermore and San Diego using nearly monochromatic γ-rays ($\Delta E_\gamma \sim 300$ keV for $E_\gamma = 15$ MeV) obtained by positron annihilation-in-flight In these experiments, the total yield of neutrons is measured as a function of E_γ in the range ~ 10 to 30 MeV.

3.1. (γ, n) REACTIONS

At the present time, there are two methods being used to study (γ, n) reactions using electron linacs. They are, firstly, measurements of the energy spectra of photoneutrons by the nanosec. time-of-flight method (for incident bremsstrahlung spectra) and, secondly, the measurement of the total yield of neutrons resulting from an incident photon spectrum which is nearly monochromatic. Betatron groups are currently limited to alternative techniques since the circular machines do not produce the intense photon fluxes required to carry out the two experiments mentioned above.

3.2. NANOSECOND TIME-OF-FLIGHT TECHNIQUES

At the first accelerator conference of this series (1958) Phil Sargent described the pioneering nanosecond time-of-flight experiment using the M.I.T. 18 MeV electron linear accelerator[92]. This machine produces electron pulses of $\lesssim 4$ nsec. duration at peak analysed currents of about 100 mA. Their early measurements of the energy spectra of photoneutrons from the reactions $Be^9(\gamma, n)$, $C^{13}(\gamma, n)$ and $Au(\gamma, n)$ etc. demonstrated the potentialities of this technique[93]. Unfortunately the energy < 18 MeV is not sufficiently high to extend their measurements into the interesting giant resonance

region of many light nuclei (typically 18 to 30 MeV).

In 1961, we set up a nanosecond time-of-flight system (10 nsec., 0.1 amp) on the Harwell machine working in the energy range $15 < E_\gamma < 32$ MeV[20]). The machine and fast pulsing system have subsequently been improved and the present performance is

pulse duration: 5 nsec,
peak analysed current: $\frac{3}{4}$ A,
repetition rate: 400 p.p.s,
energy range: $15 < E_\gamma < 39$ MeV.

The general layout of the Harwell experiment is shown in fig. 13. We are using a diode electron gun with an electrostatic beam deflection system for modulating the pulse duration. This gun is designed and built by

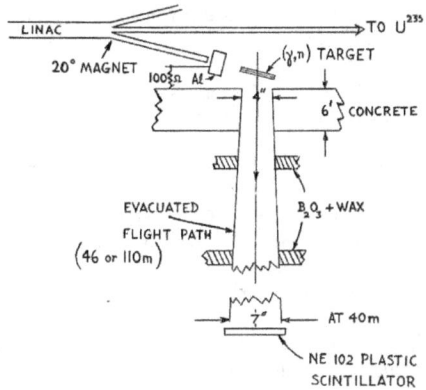

Fig. 13. General layout of the Harwell nsec neutron time-of-flight system[20]).

Vickers Research Ltd. Weybridge, England. The electron injection energy is ~ 50 keV at peak currents ~ 2 A.

Using a flight path of 40 metres and an overall time resolution of 6 nsec a time-of-flight resolution of 0.15 nsec/m is achieved*. This corresponds to an energy uncertainty of only 3.7 keV for a 1 MeV neutron increasing to 30 keV for a 4 MeV neutron. Thus, in measuring energy spectra of photoneutrons from light nuclei, (threshold 16 MeV, say), we are studying the levels at excitation energies from 17 to 25 MeV with resolutions from 4 to 100 keV respectively (ΔE_n varies as $E_n^{\frac{3}{2}}$ in a time-of-flight spectrometer).

As an example of the power of this method fig. 14 shows the observed spectrum of photoneutrons from the reaction $O^{16}(\gamma, n)O^{15}$ for an incident bremsstrahlung

---

* Six days ago the flight path was extended to 110 metres giving a resolution of 50 pico sec/m.

92) C. P. Sargent, Proc. Accelerator Conf. Cambridge, Mass. (1958).

93) Bertozzi, Paolini and Sargent, Phys. Rev. (L) **110** (1958) 790.

216   F. W. K. FIRK

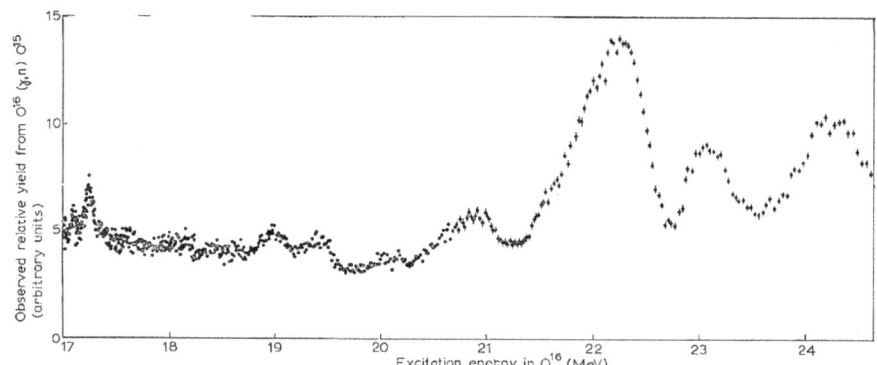

Fig. 14. The observed relative yield of photoneutrons from the reaction $O^{16}(\gamma,n)O^{15}$ when irradiated with 25 MeV bremsstrahlung
The resolution is 0.15 nsec/m [95].

energy of 25 MeV[94,95]). The energy levels in the region of the giant resonance are well resolved. Here, the timing channels are 2.3 nsec duration and the flight path length 37 metres. (It is of interest to compare this measurement with the earlier work of ref.[96].) I shall return to a discussion of the $O^{16}(\gamma,n)$ work later when describing some recent results from Livermore.

Probably the most striking resemblance between the present (γ,n) measurements and the usual concept of nuclear spectroscopy is shown in fig. 15. This depicts the $S^{32}(\gamma,n)S^{31}$ spectrum for two incident bremsstrahlung energies[95]. In the region $E_{exc} < 20$ MeV we are

94) Firk and Lokan, Phys. Rev. Letters 8 (1962) 321.
95) F. W. K. Firk, Nuclear Phys. 52 (1964) 537.
96) Milone and Rubino, Nuovo Cimento 13 (1959) 1035.

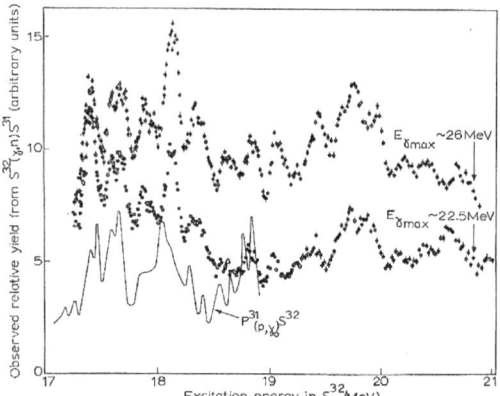

Fig. 15. A comparison between the observed yield of neutrons from the reaction $S^{32}(\gamma,n)S^{31}$ and the $P^{31}(p,\gamma_0)S^{32}$ data[89,95].

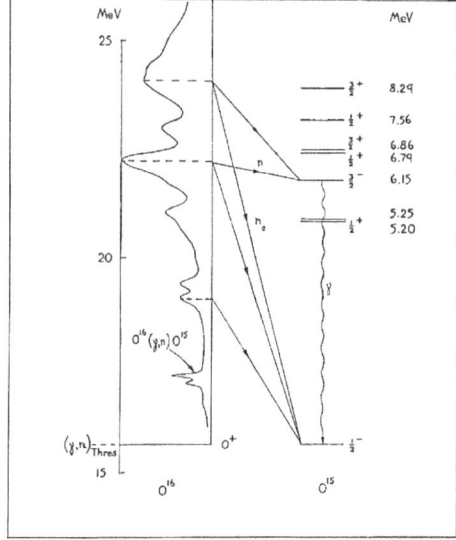

ENERGY LEVELS IN O^{16} AND O^{15}

Fig. 16. Energy levels in O^{16} excited by photons and the level scheme in O^{15} showing typical modes of neutron decay.

able to make a direct comparison with the inverse reaction $P^{31}(p,\gamma_0)S^{32}$ measured on the Harwell tandem[89]. There is remarkably close agreement between the energies and relative strengths of many of the levels. I am looking forward to results from three-stage tandems which will enable the (p,γ) work to go to considerably higher excitation energies. We have already made (γ,n) measurements up to 32 MeV in S^{32}.

There is one point which may be puzzling a number

of people, that is: why do we observe a line spectrum of neutrons when we irradiate a target with a continuous (bremsstrahlung) photon spectrum? The reasons are explained most easily by looking at fig. 16, which shows the energy levels in O^{16} and the residual nucleus O^{15}. When a nucleus absorbs photons it does so into discrete energy levels and is, essentially, a fine monochromator. If, therefore, the resolving power of the neutron spectrometer is sufficient to separate the neutron groups emitted from the (γ, n) target and, furthermore, there are few levels in the residual nucleus which can be populated by the photoneutrons then one will observe a line spectrum of neutrons. The two cases used as examples of the method are particularly favourable: 0^+ ground states going to 1^- states by E1 absorption (which is the predominant mode in the giant "dipole" resonance region) and only the ground state, or at most, a few excited states available in the residual nucleus. To establish the ground state cross section, it is necessary to choose the bremsstrahlung energy to be just below the first excited state in the residual nucleus. With these "positive" experiments, however, in which definite levels are observed, relatively crude "photon-difference" methods may be used effectively provided the level spacing in the residual nucleus is wide. Needless to say, it is a great help in interpreting the (γ, n) data to be able to compare it with high resolution (p, γ_0) measurements.

Earlier in this talk I mentioned the work which has been carried on in slow neutron spectroscopy in an effort to study the distribution of partial radiation widths. In some cases, it is possible to study the distribution of ground state radiation widths by means of (γ, n) reactions. The area under an isolated level in a (γ, n) yield curve is proportional to $g\Gamma_{\gamma 0}(\Gamma_n/\Gamma)$. For many light nuclei the ratio (Γ_n/Γ) is apparently constant (comparing (γ, n) with (p, γ_0) etc.) so that fluctuations in the areas of the (γ, n_0) yield curves represent fluctuations in the ground state radiation width $\Gamma_{\gamma 0}$. This

technique has been applied to 26 levels in $S^{32}(\gamma, n_0)S^{31}$ and the distribution $\Gamma_{\gamma 0}/\bar{\Gamma}_{\gamma 0}$ compared with a Porter-Thomas and an exponential distribution. (here, $\Gamma_{\gamma 0} = \Gamma_{\gamma 0}/E_\gamma^3$). It is found that $v \approx 1$ is not at all inconsistent with the data[95]).

Two other linac groups have recently reported (γ, n) measurements using the nanosecond time-of-flight method. At R.P.I., Yergin et al.[21]) have achieved a 10 nsec burst, > 1 A peak current and are using a 100 metre flight path. The photon energy is continuously variable from about 20 to 60 MeV. In Japan (Kimura et al.[22]), the burst width obtained is < 40 nsec and a flight path length of 50 metres is used, $(E_\gamma \lesssim 25$ MeV). We may, therefore, expect many more interesting results in this field during the next few months.

3.3 MONOCHROMATIC γ-RAY TECHNIQUES

The positron annihilation-in-flight method for producing nearly monochromatic γ-rays has been firmly established in several laboratories during the last few years[23-25]). Measurements of the absolute (γ, n) cross sections of many nuclei have demonstrated the importance of this new method in photonuclear reaction studies.

The three groups now making such measurements all achieve about the same experimental performance from their machines. Typical figures are

mean analysed positron current $\quad \sim 5 \times 10^{-12}$ A,
energy resolution at 20 MeV $\quad \Delta E_\gamma/E_\gamma \sim 2\frac{1}{2}\%$,
energy range $\quad 10 < E_\gamma < 30$ MeV.

There are two different approaches to the problem of producing usable intensities of high energy positrons. At Saclay, Tzara and co-workers[23]) accelerate electrons to the full machine energy and then produce electron-positron pairs in a thick tungsten target. The positrons are selected using a magnet system which also serves as an energy analyser. The positrons are annihilated in a Be or LiH target in which photons are generated into a narrow solid angle. The energy spread is due to a

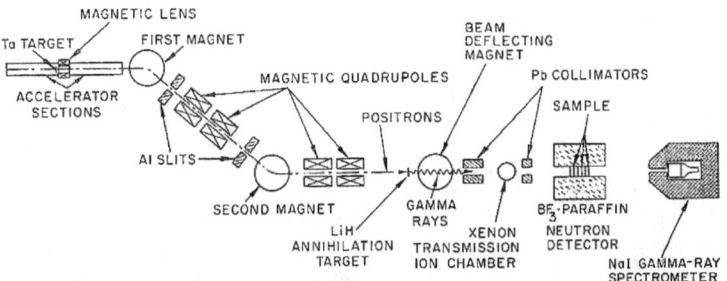

Fig. 17. Layout of the positron annihilation-in-flight experiment at Livermore[24]).

VI. EXPERIMENTAL TECHNIQUES

84

218   F. W. K. FIRK

number of factors such as beam divergence of the incident positron beam, multiple scattering of positrons in the target and the finite solid angle subtended by the actual (γ,n) target. In order to obtain a reasonable flux of γ-rays it is necessary to compromise between target thickness, and solid angles and, with presently available beam powers from accelerators, the overall resolution is then about $2\frac{1}{2}\%$ for a 15 MeV γ-ray.

At Livermore[24] and San Diego[25] an ingeneous method is used in which electrons are accelerated to an energy of about 10 MeV in the early part of the machine. A thick tungsten target is placed in the accelerator and electron-positron pairs are produced. By changing the phase of the r.f. in the following accelerator sections only positrons are accelerated to the full energy. These positrons are then magnetically analysed and allowed to annihilate in a LiH target to produce γ-rays of well-defined energy. The layout of the experiment at Livermore is shown in fig. 17. The γ-rays irradiate a suitable (γ,n) target which is surrounded by

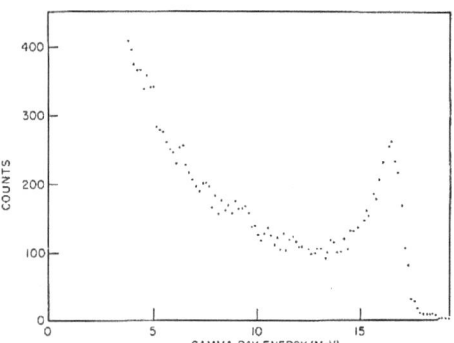

Fig. 18. Typical spectrum of γ-rays from positron annihilation as observed in a NaI(Tl) crystal. The bremsstrahlung background when electrons are accelerated appears identical except the 15 MeV peak is no longer present[24].

a neutron detector which has an efficiency for detecting neutrons which is reasonably constant for $1 < E_n < 10$ MeV. By using many BF_3 counters in the detector array it is possible to measure the photoneutron yield for the $(\gamma,2n)$ reaction by comparing the singles with the coincidence counting rates. Backgrounds are determined by changing the r.f. phase and the magnet polarities thereby accelerating electrons; the conditions for producing bremsstrahlung backgrounds are identical with the positron case (see fig. 18).

Recent examples of this powerful method are shown in fig. 19 [97] and fig. 20. The $O^{16}(\gamma,n)O^{15}$ measurements[98] are interesting since a direct comparison is

possible with the Harwell $O^{16}(\gamma,n_0)O^{15}$ yield curve. Table 3 lists the two sets of results and compares them with the calculations of Elliott and Flowers[99] and Brown et al.[100]. There are notable discrepancies since both experiments indicate that at least 25% of the yield below 25 MeV comes from states other than the two at

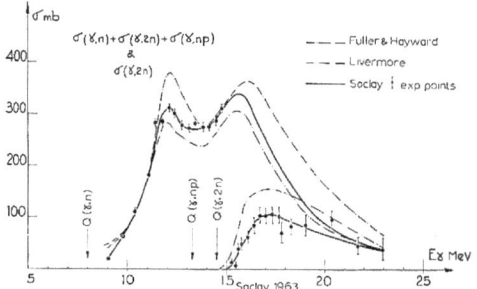

Fig. 19. The (γ,n) and $(\gamma,2n)$ cross sections of Ho as measured at Saclay[97] and Livermore[98].

22.3 and 24.3 MeV. A recent model of photo-absorption in O^{16} has been proposed by Boeker[101] in which he suggests that the particle-hole states may be coupled to 0^+ or 2^+ surface vibrations thereby causing a splitting of each "state". Such a mechanism certainly helps explain the observed data.

Although the resolution of the monochromatic γ-ray technique is currently a few hundred keV there is a great deal of new and important data to be obtained using it as the two examples clearly show.

Useful photon fluxes with resolutions of about 1%

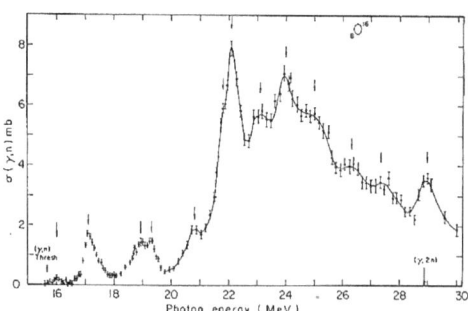

Fig. 20. The $O^{16}(\gamma,n)O^{15}$ cross section obtained with the Livermore photon monochromator[98].

97) C. Tzara, private communication (1963).
98) Bramblett and Fultz, private communication (1963).
99) Elliott and Flowers, Proc. Roy. Soc. (Lond) **242** (1957) 57.
100) Brown, Castillejo and Evans, Nuclear Phys. **22** (1961) 1.
101) E. Boeker, Thesis, Amsterdam (1963).

TABLE 3

A comparison between the (γ, n_0) yields[95], the total (γ, n) yields[98] and the theoretical predictions of references[99] and [100]. Only the relative yields up to 25 MeV are considered.

Harwell Experiment[95] (Ground state yield)		Bramblett[98] (Total neutron yield)		Brown et al.[100]		Elliott & Flowers[99]	
E(MeV)	Rel. int. (%)	E(MeV)	Rel. int. (%)	E(MeV)	Rel. int. (%)	E(MeV)	Rel. int. (%)
17.10	0.5						
17.25	2.0	17.3	3	17.6	1	17.3	1
19.0	4.0	19.0	3				
19.4	4.0	19.3	3				
				20.0	1	20.4	0
20.1	1.5						
20.9	8.5	20.8	6				
21.6	2.5	21.7	1				
22.1 }22.3 }	40	22.1	38	22.2	±68	22.6	67
23.1	11	23.1	16				
24.1 }24.3 }	26	24.1	30	25.0	29	25.2	32

seem quite feasible and will, no doubt, be obtained within the next year.

3.4. (γ, p) REACTIONS

The first high resolution experiments on (γ, p) reactions* were carried out at Stanford[102] using a magnetic proton spectrometer with a resolution of about 2%. This experiment was among the first to show detailed fine structure in the giant resonance regions of O^{16} and Ne^{20}.

Recently, the 45 MeV Yale electron linac has been used to measure (γ, p) reactions using solid state counters to determine the proton energies[103]. Severe background difficulties arise due to the intense γ-flash from the accelerator which may produce pulses in the counters which have pulse heights indistinguishable from those due to protons or which cause complete paralysis of the counter system. The Yale group have largely overcome these difficulties by extracting the protons from the (γ, p) target via a magnetic lens. The protons are then detected in a well shielded room which is unaffected by the forward moving photons from the γ-flash. Resolutions of < 1% are attainable with this technique and the use of thick Li-drifted counters enable the entire giant resonance region to be explored.

The successes of the Yale Group will obviously en-

courage other linac groups to take advantage of the high resolutions which may be achieved. We may, therefore, look forward to precise measurements of (γ, p), (γ, d) and (γ, α) reactions in the near future.

4. Conclusions

The results obtained using electron linacs in both slow neutron and photonuclear reactions fully justify their current popularity. This talk has emphasized the fundamental nuclear physics processes which are being investigated with their help. I have not mentioned the obvious applications of the slow neutron work to the important task of obtaining data for nuclear reactor programmes as this topic has been discussed at numerous conferences.

The combination of results from electron linacs with those from other charged particle accelerators will continue to add considerably to our understanding of the nucleus.

Acknowledgements

It is a pleasure to acknowledge the help of many colleagues who have kindly supplied recent work much of which is unpublished. They include C. K. Bockelman, F. D. Brooks, J. E. Draper, S. C. Fultz, E. Haddad, G. D. James, M. Kimura, A. Michaudon, M. C. Moxon and F. Netter.

* The reaction used was (e,e′p) which outwardly behaves as a (γ, p) reaction.

102) Dodge and Barber, Phys. Rev. 127 (1962 1746.
103) O'Connell and Bockelman, private communication (1963).

LOW-ENERGY PHOTONUCLEAR REACTIONS

F. W. K. FIRK

Physics Department, Yale University, New Haven, Connecticut

CONTENTS

INTRODUCTION

One of the best-understood theories in the field of nuclear reactions is that which describes the interaction between photons and nuclei (1, 2). The energy dependence of the absorption process exhibits a broad resonance that is observed in all nuclei at energies ranging from 4 to 30 MeV (3, 4). The most important characteristics of the resonance are a consequence of the electric

40   FIRK

dipole nature of the interaction which dominates the absorption process at these energies (5).

In medium and heavy nuclei, the gross properties of the resonance (such as the dependence on mass number of the resonance energy and of the integrated absorption cross section) are reasonably well understood (6–8). In these cases, the level densities are usually so high that the observations only provide information of a statistical nature (an exception being the threshold photoneutron work discussed on page 65).

In light nuclei, however, the densities of levels excited by E1 photons are frequently so low that the quantum numbers associated with individual levels can, in principle, be determined. This feature has resulted in a considerable overlap between recent experimental studies of photonuclear reactions and other studies of nuclear spectroscopy [particularly those associated with proton and neutron induced reactions (9, 10)]. Moreover, theoretical descriptions of photonuclear reactions are frequently given in terms of single- or few-particle excitations coupled to the continuum states (11–13). Such descriptions have much in common with other branches of nuclear-structure and nuclear-reaction theory. Previous reviews in this series have followed the development of the field from its beginning in the late 1940s until early 1965 (14–17). The present review will concentrate, for the most part, on those developments reported between 1965 and early 1970 which have contributed to the common cause.

The photon energies of interest will be below about 30 MeV. Topics, such as the quasideuteron effect (15), that are associated with the absorption of higher energy photons will therefore be omitted. Discussions of the photodisintegration of the mass 2, 3, and 4 nuclei will also be omitted. Experimental and theoretical work in these special nuclei is in a fluid state and a review seems inappropriate.

SOME GENERAL PROPERTIES OF THE DIPOLE STATE
A-DEPENDENCE OF THE RESONANCE ENERGY IN HEAVY NUCLEI

Estimates of the energy of the dipole state in nuclei with $A \gtrsim 100$ obtained from a harmonic-oscillator model give values of $E_{res}^{(h.o.)} = \hbar\omega_0 = 41A^{-1/3}$ MeV (6). This is to be compared with the observed values of $E_{res}^{(obs)} \simeq 80\ A^{-1/3}$ MeV (see Table 1). Recently, Bohr & Mottelson (8) proposed a model in which the effects of nuclear vibrational modes are specifically included. The resonance energies are then found to be in much better agreement with the observations (see Figure 1a). Such vibrations involve a coherent motion of nucleons which results in density variations with respect to the equilibrium density. The resulting variation in nuclear potential δV is related to the vibrational amplitude α as follows:

$$\delta V \simeq \chi F(x)\alpha \quad \text{(to first order in } \alpha) \qquad\qquad 1.$$

where $F(x)$ describes the dependence of the potential on the space, spin, and

TABLE 1. Some general properties of the giant dipole states in heavy nuclei[a]

Nucleus	$\sigma_{\text{int}}^{(\text{obs})}$ (MeV-b)	$E_{\gamma\text{max}}$ (MeV)	$0.06NZ/A$ (MeV-b)	α (MeV)	K[b] (MeV)
[89]Y	1.04	28.0	1.31	75.0	23.0
[90]Zr	1.06	28.0	1.33	75.4	23.2
[91]Zr	1.08	30.0	1.35	74.6	22.7
[92]Zr	1.10	28.0	1.36	73.6	22.2
[94]Zr	1.04	30.0	1.38	73.6	22.6
[107]Ag	1.35	29.5	1.58	76.4	24.2
[115]In	1.90	31.1	1.69	76.0	24.1
[116]Sn	1.67	29.4	1.71	76.5	24.1
[117]Sn	1.94	31.1	1.72	76.6	24.4
[118]Sn	1.90	30.0	1.73	76.5	24.4
[119]Sn	2.08	31.1	1.74	76.4	24.4
[120]Sn	2.09	29.9	1.75	76.0	24.2
[124]Sn	2.08	31.1	1.79	75.7	24.2
[133]Cs	1.98	29.5	1.93	77.0	25.0
[141]Pr	2.10	30.0	2.06	78.9	26.0
[153]Eu	2.28	28.9	2.22	78.3	25.9
[159]Tb	2.30	28.0	2.31	78.7	26.2
[160]Gd	2.53	29.5	2.30	79.9	27.3
[165]Ho	2.51	28.9	2.39	80.2	27.2
[181]Ta	2.19	24.6	2.61	80.8	27.3
[186]W	3.00	28.6	2.67	80.5	27.3
[197]Au	2.97	25.0	2.87	80.6	27.9
[206]Pb	2.78	28.0	2.96	80.9	27.3
[207]Pb	2.65	28.0	2.97	80.5	27.6
[208]Pb	2.91	28.0	2.98	80.6	27.9
[209]Bi	2.93	28.0	3.00	80.2	27.6

[a] These results are taken from the work of Fultz and colleagues (21–26) and from references contained therein. The errors on the observed integrated cross sections, $\sigma_{\text{int}}^{(\text{obs})}$, are generally less than 10%.

[b] The nuclear-symmetry energy K is determined from the relationship (32):

$$E_{\text{res}} = \frac{\hbar k}{A}\left\{\frac{8KNZ}{M}\left[1 - \left(\frac{\Gamma}{2E_{\text{res}}}\right)^2\right]\right\}^{1/2}$$

where $kR = 2.08$ for spherical nuclei and R is the nuclear radius.

isospin variables of the nucleons and χ is the coupling constant between the density variations and the field generated by the vibrational mode.

The giant dipole states of nuclei involve isovector modes in which the neutrons and protons move in antiphase. The nuclear-symmetry potential is therefore used to estimate the coupling between the density and field oscillations in these modes. The nuclear potential may be written (18):

42   FIRK

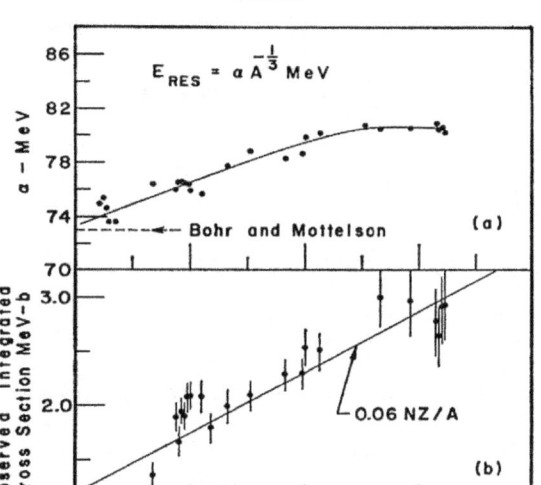

FIGURE 1. (a) The observed dependence of the quantity $\alpha = E_{res}A^{1/3}$ (MeV) (21–26). The curve has no quantitative significance. In the case of deformed nuclei, a mean value $E_{res} = (E_{lo} + 2E_{hi})/3$ is used in which E_{lo} and E_{hi} are the lower and upper resonance energies. The value of $\alpha \simeq 73$ MeV estimated by Bohr & Mottelson (8) using Equation 8 is indicated.

(b) The observed integrated total photoneutron cross sections for heavy nuclei (21–26). The energy limits of the integrals are from threshold to approximately 30 MeV. (See Table 1 for details.) The value of $0.06NZ/A$ (MeV-b) obtained from the classical dipole sum rule is also shown.

$$V = V_0 + (1/2A)V_1 t_z (N - Z) \qquad\qquad 2.$$

$$= V_0 + (1/2)V_1 t_z (\rho_1/\rho_0) \qquad\qquad 3.$$

where the second term describes the potential produced by a small isovector density $\rho_1 = \rho_n - \rho_p$ superimposed on the isoscalar density $\rho_0 = \rho_n + \rho_p$. Lane (19) gives a value of $V_1 \simeq 100$ MeV. If the density variation of the isovector mode is of the form:

$$\delta\rho_1 = f(r)\alpha \qquad\qquad 4.$$

the associated potential is expected to be of the form:

$$\delta V \simeq (V_1/2\rho_0)f(r)t_z\alpha \qquad\qquad 5.$$

In the dipole mode, $f(r)$ is assumed to be proportional to r.

The amplitude α is normalized as follows:

$$\alpha = (2/A) \sum_{k=1}^{A} (x_k t_{zk})$$

$$= (1/A) \int \delta \rho_1 x d^3 r \qquad \qquad 6.$$

Considering motion along the x axis:

$$f(\mathbf{r}) = (\rho_0 / \langle x^2 \rangle) x$$

where $\langle x^2 \rangle$ is the mean value of x^2 for a particle in the nucleus. It then follows that

$$\delta V \simeq (V_1/2\rho_0) f(\mathbf{r}) t_z \alpha = (2/A) \chi x t_z \alpha \qquad \qquad 7.$$

in which

$$\chi = (5A/4R^2) V_1$$

In the dipole state, the single-particle excitations cluster around the energy of the harmonic oscillator. The coupling with the vibrational modes then leads to an increase in the resonance energy which is now approximately given by:

$$E_{res} \simeq E_{res}^{(h.o.)} (1 + \chi / M A \omega_0^2)^{1/2}$$

$$\simeq 73 A^{-1/3} \text{ MeV} \qquad \qquad 8.$$

assuming $V_1 = 100$ MeV and $R = 1.2 A^{1/3}$ F. This value is shown in Figure 1a: it is close to the observed value for nuclei in the region $A \sim 100$.

INTEGRATED ABSORPTION CROSS SECTIONS

Comparisons between the observed total photon absorption cross sections and the predictions of the theoretical dipole sum rule can provide important information on the E1 nature of the absorption process and on the effect of exchange terms in the nuclear potential (20).

Most observed integrated cross sections have been obtained from measurements of the total (γ,n) cross sections. This is a valid procedure in heavy nuclei ($A \gtrsim 100$, say) in which the proton decay of the dipole state is severely restricted by the Coulomb barrier. Early measurements of the (γ,n) cross sections suffered from two main difficulties: the use of photon difference methods with the attendant uncertainties resulting from unfolding bremsstrahlung spectra; and the inability to obtain both the single and multiple neutron emission cross sections.

Over the past few years, these difficulties have been overcome in an extensive series of measurements carried out by Fultz and his colleagues (21–26) using the Livermore photon monochromator. In these measurements, nearly monochromatic photons are produced by the annihilation-in-flight of

positrons (Tzara 27, Schuhl & Tzara 28, Jupiter et al 29). The photoneutrons are detected in a high efficiency 4π neutron counter (21, 30) and the
single, double, and triple neutron cross sections determined by observing the
appropriate coincidences during the "neutron-slowing-down time" of the
detector. A precise and self-consistent set of measurements of the total photoneutron cross sections of many nuclei ranging from ^{3}He to ^{235}U is now
available. The observed integrated cross sections from threshold to approximately 30 MeV are presented in Table 1 for the heavy nuclei ($90 < A < 210$).
The results are also shown in Figure 1b in which they are compared with the
value of $0.06NZ/A$ (MeV-b) given by the classical dipole sum rule (no exchange forces).

Several comments can be made regarding these results: the measured
integrated cross sections generally exceed the classical dipole sum for nuclei
with values of $100 \lesssim A \lesssim 200$ even though the integral is limited to 30 MeV.
If the measurements were extended to higher energies (to the meson threshold, at least) the observations could clearly exceed the predicted values by
sizable ($\gg 10\%$) amounts. The Livermore group conclude, however, that the
integrated cross sections are in reasonable agreement with the classical dipole
sum quoting little ($\lesssim 10\%$) or no correction due to exchange effects. This
conclusion is primarily a consequence of their method of data analysis in
which a Lorentz curve (or curves in deformed nuclei) is fitted to the observed
resonance (usually the lower-energy side) and is then extrapolated to an infinite energy. This would seem to be a procedure that requires a more rigorous justification than has been given so far.

Additional measurements are therefore required at energies above 30
MeV. Furthermore, it is important to establish the purity of the E1 absorption process at such high energies: both angular distribution and polarization
measurements will be required before definite conclusions can be drawn
from the results (see page 59).

EFFECTS OF NUCLEAR DEFORMATION

*Determination of the nuclear eccentricity, radius parameter, and quadrupole
moment.*—Okamoto (31) and Danos (32) developed the hydrodynamic
model (Steinwedel & Jensen 33) in order to include the effects of nuclear deformation. They showed that the dipole state should be split into two components corresponding to excitations along and perpendicular to the nuclear-
symmetry axis. Experimental evidence for such splitting was first obtained
12 years ago (34, 35).

Danos (32) derived several important relationships namely:

$$E_{lo}/E_{hi} = 0.911\eta + 0.089 \qquad\qquad 9.$$

where E_{lo} and E_{hi} are the lower and higher energies of the two components
of the dipole state, respectively, and η is the ratio of the lengths of the major
to the minor axis. The eccentricity of the nucleus, ϵ, is related to η as follows:

$$\epsilon = (\eta^2 - 1)\eta^{-2/3}$$
$$= (b^2 - a^2)/R^2$$

10.

where a and b are the semimajor and semiminor axes, and R is the radius of the sphere with a volume equal to that of the ellipsoid:

$$R^3 = a^2 b = (r_0 A^{1/3})^3$$

11.

where r_0 is the nuclear radius parameter.

Berman et al (25) have recently reported measurements of the total photoneutron cross sections of six strongly deformed nuclei (^{153}Eu, ^{159}Tb, ^{160}Gd, ^{165}Ho, ^{181}Ta, and ^{186}W) using the Livermore photon monochromator and its associated neutron detector. A typical result is shown in Figure 2, which shows the total and partial photoneutron cross sections of ^{160}Gd in the region of the dipole state. The splitting is very pronounced—note the importance of measuring the $(\gamma,2n)$ cross section with precision. The position of the $(\gamma,2n)$ threshold happens to occur between the two energies E_{lo} and E_{hi} so that the decay of the upper resonance is associated, primarily, with the $(\gamma,2n)$ process. By fitting the data with two Lorentz curves, it is possible to locate the resonance energies with considerable accuracy (an uncertainty of <100 keV at 15 MeV). Values of η and ϵ are thereby deduced using Equations 9 and 10.

The quadrupole moment is also simply related to R and ϵ:

$$Q_0 = \tfrac{2}{5}ZR^2\epsilon$$

12.

This quantity is obtained, independently, from Coulomb excitation studies (and also from electron scattering and mu-mesic atom studies). Therefore, a combination of the values of Q_0 deduced from such studies with the values of ϵ, derived from the photoneutron experiments, yield precise values of R and hence r_0 (the nuclear radius parameter). The results are listed in Table 2.

A mean value $\langle r_0 \rangle = 1.26 \pm 0.02$ F is obtained from the six values of r_0 so that "optimum" values of $Q_0^{(opt)}$, the quadrupole moment, derived from Equation 12 using the measured values of ϵ and a constant value of $\langle r_0 \rangle$, can be found. These values are also given in Table 2.

Measurements of the photoneutron cross section from aligned ^{165}Ho.—A stringent test of the model for deformed nuclei is provided by measurements of the total photon absorption cross section as a function of the angle between the incident photon beam direction and the nuclear-symmetry axis. Kelly et al (43) have carried out the most definitive measurement of this kind to date.[1] In an elegant experiment, they measured the (γ,n), (γ,pn),

[1] An earlier experiment of Ambler, Fuller & Marshak (44) observed an anisotropy in the dipole resonances of ^{165}Ho on changing the orientation of the nuclear-symmetry axis with respect to the direction of the incident photon beam. The energy dependence of the effect was not measured, however.

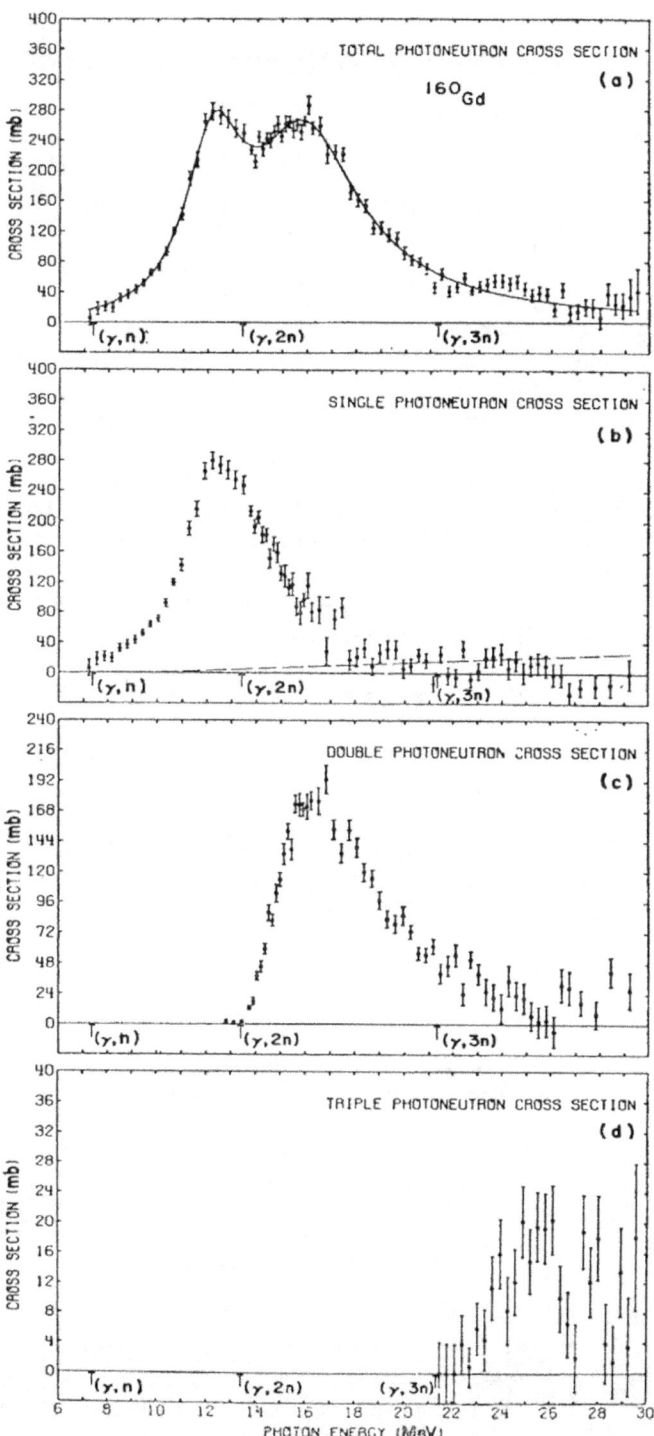

FIGURE 2. The total and partial photoneutron cross sections for ^{160}Gd (25). The splitting of the dipole state, due to nuclear deformation, is clearly seen. Note the importance of a precise measurement of the $(\gamma,2n)$ cross section.

94

TABLE 2. Nuclear eccentricities, radius parameters and quadrupole moments[a]

Nucleus	ϵ	$Q_0^{(el)}$ (b)	r_0 (F)	$Q_0^{(opt)}$ (b)	Ref. for $Q_0^{(el)}$
^{153}Eu	0.595 ± 0.015	6.99 ± 0.08	1.276 ± 0.018	6.80 ± 0.28	(36, 37)
^{159}Tb	0.598 ± 0.009	7.41 ± 0.11	1.274 ± 0.013	7.23 ± 0.26	(36)
^{160}Gd	0.645 ± 0.014	7.55 ± 0.17	1.245 ± 0.020	7.71 ± 0.30	(38)
^{165}Ho	0.604 ± 0.006	7.56 ± 0.11	1.246 ± 0.011	7.71 ± 0.26	(36)
^{181}Ta	0.433 ± 0.010	6.89 ± 0.21	1.306 ± 0.025	6.41 ± 0.26	(39, 40)
^{186}W	0.390 ± 0.006	5.96 ± 0.05	1.259 ± 0.011	5.96 ± 0.21	(38, 41, 42)

[a] From Berman et al 1969. *Phys Rev.* 185: 1576.

and $(\gamma,2n)$ cross sections (essentially the total cross section) of aligned ^{165}Ho between 10 and 21 MeV using nearly monochromatic photons. The energy dependence of the cross section was studied for the two cases in which the nuclear-symmetry axis was either parallel or perpendicular to the direction of the incident photon beam (see Figure 3).

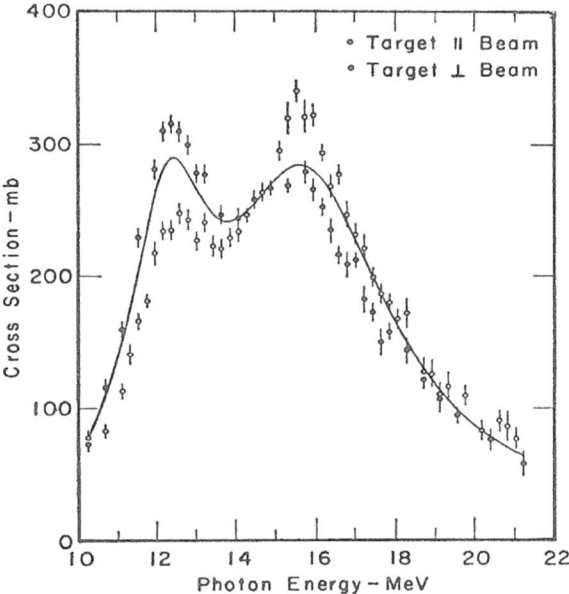

FIGURE 3. The total photoneutron cross section of ^{165}Ho aligned either parallel or perpendicular to the direction of the incident (unpolarized) photon beam. The change in the intensity of the components of the resonance, which results from a reorientation of the nuclear spin, is beautifully demonstrated (Kelly et al 43). The solid curve shows the unaligned cross section.

48   FIRK

If the projections of the spin I of the ground state $|i\rangle$ and of the spin I_j of the excited state $|j\rangle$ along the nuclear-symmetry axis are K and K_j, respectively, then the dipole states in the parallel mode have $\Delta K = 0$ and the states in the perpendicular mode have $\Delta K = \pm 1$. (Here, $\Delta K = K - K_j$.) Intrinsic cross sections may then be defined that correspond to the two modes of excitation, thus (43):

$$\sigma_{\parallel} = \delta_{\Delta K,0}(4\pi k_\gamma/3) \sum_j \mid \langle j \mid \boldsymbol{E}1 \mid i\rangle \mid^2 \frac{E_\gamma E_j \Gamma_j}{(E_j^2 - E_\gamma^2) + E_\gamma^2 \Gamma_j^2} \quad 13.$$

$$= \delta_{\Delta K,0} \, \Phi, \text{ say}$$

and

$$\sigma_\perp = \delta_{\Delta K,\pm 1} \, \Phi \qquad\qquad 14$$

where k_γ is the photon wavenumber, E_γ the photon energy, $\boldsymbol{E}1$ the electric dipole operator, E_j and Γ_j the mean energy and width of the jth resonance, respectively, and δ the Kronecker delta function. The measured intrinsic cross sections are shown in Figure 4.

THEORETICAL DEVELOPMENTS

The fine structure revealed in the dipole states of most nuclei clearly demonstrates that a complete theoretical description of such complex states is not feasible. Current theories are therefore directed towards an understanding of the observed gross and intermediate structure.

Absorption Processes

Theories of the absorption process, which describe the interaction between a photon and A nucleons to form an intermediate state of given spin and parity, are generally based upon a single-particle model.

Most approaches stem from the work of Elliott & Flowers (11) who predicted the locations and strengths of 1^- states in ^{16}O formed by absorbing E1 photons. The excited states are described as linear combinations of 1 particle–1 hole ($1p$-$1h$) states. The main part of the electric dipole strength is then found to be concentrated in a few states consisting of a coherent combination of the basis states (see also Brown & Bolsterli 46).

Use of realistic forces.—In all recent calculations, the energies of the unperturbed particle-hole states are taken from experimental observations. This may be expedient from a calculational point of view but it (unfortunately) bypasses the basic problem of calculating the energies from a microscopic theory of nuclear structure.

New work by Kuo, Blomqvist & Brown (47) calculates the energies of the dipole states in ^{208}Pb with "realistic" nuclear forces. Using the G-matrix elements of the Hamada-Johnston potential as effective interactions (48), Kuo et al (47) compute the energies and strengths of the dipole states with

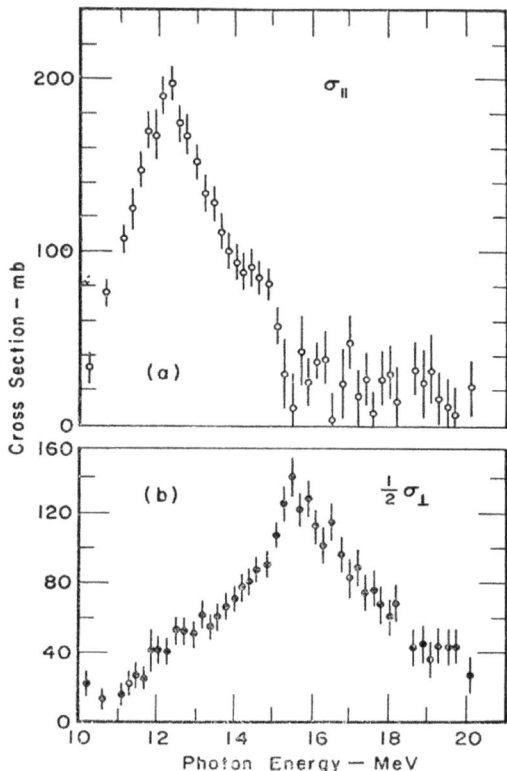

FIGURE 4. (a) The intrinsic photoneutron cross section of [165]Ho aligned parallel to the direction of the incident photon beam (43).

(b) The intrinsic photoneutron cross section of [165]Ho aligned perpendicular to the direction of the incident photon beam. The departure of these curves from Lorentz shapes indicates the need for a more elaborate hydrodynamic theory (43, 45).

the same unperturbed shell-model states as Gillet et al (49). The results are found to be almost 3 MeV lower in energy than the observed resonance energy of 13.9 MeV. Since the unperturbed energies are centered at approximately 7 MeV, the effective interaction used by Kuo et al (47) only increases the resonance energy by 60% of the required amount. This is in contrast to calculations in lighter nuclei (50) where the G-matrix elements worked well in reproducing the necessary energy shifts. Furthermore, it had been anticipated that the inclusion of "core-polarization" effects in the calculation would have a considerable effect on increasing the calculated energy of the dipole state in [208]Pb. However, the results show a rather small effect: Kuo et al (47) suggest that this may reflect the fact that the dipole state of [208]Pb is not as "collective" as expected in which case the spatial correlations of the particle and hole are not particularly strong. (The inclusion of such

$3p$-$1h$ processes required the calculation of an additional 175,000 G-matrix elements.) The authors feel that the problem should also be recalculated using a Woods-Saxon potential in which the neutron and proton excitations are considered separately.

Another possibility, which is in the spirit of a calculation of Kamimura et al (51), is that the dipole oscillation in ^{208}Pb is coupled to the octupole vibration associated with the 3^- excited state at 2.6 MeV. If the coupling is sufficiently strong, then an appreciable fraction of the giant resonance should occur $\sim$2.6 MeV above the value calculated by Kuo et al (47).

Open-shell calculations.—Shell-model calculations based upon the random-phase approximation are not readily applicable to nuclei other than those with doubly closed shells (or one particle or hole removed). Calculations in a nucleus, such as ^{12}C, therefore require a somewhat modified approach. The equations-of-motion method, as developed by Rowe (52), extends the random-phase approximation method so that it becomes suitable for handling calculations with many particles or holes in the target nucleus. A specific case is ^{12}C, in which the energies and strengths of the dipole states have been calculated by treating the target as an open-shell nucleus. Before discussing the calculation, some mention of earlier calculations of 1^- states in ^{12}C provides necessary background information. In 1962, Nilsson, Sawicki & Glendenning (53) used a deformed potential to account for the observed gross splitting of the dipole state into two parts centered at 22.5 and 25.5 MeV. The 1^- states split into groups that were identified with the $K=0$ and $K=1$ transitions. Using a reasonable value of the nuclear deformation parameter, a splitting of 3 MeV was obtained. Whether or not the observed splitting is associated with a deformation effect has not been experimentally established. (A convincing test would be provided by carrying out photon scattering experiments in which the γ-ray decay schemes were measured and compared with the predictions of a Nilsson-type calculation.)

An alternative model is that of Kamimura, Ikeda & Arima (51) who explain the considerable strength at 25.5 MeV by treating the ^{12}C ground state as a mixed $(0^+ \otimes 2^+)$ configuration. Here, the admixture arises from coupling the first 2^+ excited state in ^{12}C at 4.43 MeV to the ground state. Drechsel, Seaborn & Greiner (54) carried out a related calculation using the macroscopic approach of the dynamic collective model (45). Both calculations thereby reproduce the observed splitting of the dipole state since they reflect the energy difference between the ground- and first excited (2^+) state of the target nucleus.

Rowe & Wong (55) have calculated the $J^\pi = 1^-$, $T = 1$ states in ^{12}C, using an intermediate coupling model of the ground state rather than a closed $1p_{3/2}$ shell as used in earlier calculations (see Figure 5a). It is found that the gross features of the dipole states in ^{12}C are more readily accounted for. In particular, considerable strength is obtained in the region 25–26 MeV. The results of the calculation are shown in Figure 5b. In another version of the

98

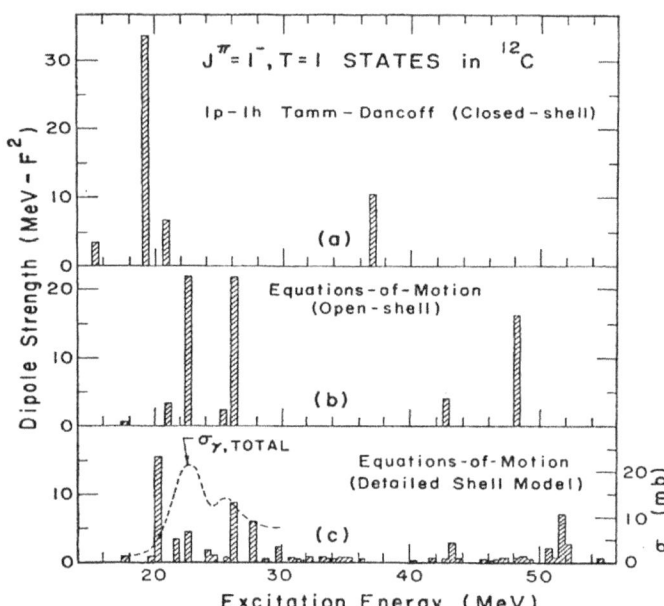

FIGURE 5. A comparison between the positions and relative strengths of 1⁻ states in ¹²C calculated using:

(a) the closed-shell Tamm-Dancoff method
(b) an open-shell equations-of-motion method
(c) a detailed shell-model treatment within the equations-of-motion framework (Wong & Rowe 56).

calculation, Wong & Rowe (56) use a shell-model approach in which the $(1s)^{-1}(1p)^9$ and $(1p)^7(2s1d)^1$ configurations are taken for the model space. A total of 141 states are considered (including, of course, the $1p$-$1h$ states of the Tamm-Dancoff approximation shown in Figure 5a). The dipole strength is now distributed over a wide range ($\sim$25 MeV) as shown in Figure 5c. Although the splitting of the dipole state is obtained, the magnitude of the effect is too large ($\sim$4.5 MeV instead of the observed value of 3 MeV). If, as Wong & Rowe (56) suggest, the splitting is a shell rather than a collective effect, it would seem difficult to reconcile their estimated splitting with the observed value of 3 MeV. The question of shell- or collective-splitting of the dipole state in ¹²C is not yet settled.

DECAY PROCESSES

Continuum effects.—The calculation of the particle decay widths of the dipole states presents some formidable problems, which can only be overcome by making simplifying assumptions concerning the reaction mechanism. The most successful approach, so far, has been the coupled-channels method of Buck & Hill (12) who calculated the cross sections for the inverse reactions ^{15}N(p,γ)^{16}O and ^{15}O(n,γ)^{16}O. The calculation was carried out within a $1p$-$1h$ basis and used a Woods-Saxon potential with an arbitrary adjustment of the imaginary part to take account of the reaction channels that were not treated explicitly. The particle widths were found to be in reasonable agreement with the gross features of the observations (86). Angular distributions and polarizations were also calculated and found to be in modest agreement with experiment (see p. 61). Similar calculations have been reported by Saruis & Marangoni (57) and a related calculation has been reported by Beres & MacDonald (58).

The eigenchannel method of Danos & Greiner (59) has been applied to the ^{16}O problem by Wahsweiler et al (60). The initial calculations encountered some difficulties, since resolved (61): the complete results of these latest calculations have not yet been reported.

The R-matrix approach has also been used to estimate the particle widths in ^{12}C. The early work of Boeker (62) had little success in obtaining reasonable values of the widths. A recent detailed calculation by Brassard (63), however, obtains good agreement with the gross features of the ground-state cross section (in both magnitude and energy dependence) up to 30 MeV. Furthermore, the angular distributions are also well reproduced.

Boson broadening of the dipole states.—The model proposed by Duke, Malik & Firk (64) attributes the broadening of structure, observed in photonuclear reactions in light nuclei, to the coupling of single-particle transitions (between fermion levels) to the high-lying collective (boson) excitations of the core. The Hamiltonian may then be written:

$$H = H_0 + H_T$$

where

$$H_0 = \sum_\lambda \left\{ C_\lambda^+ C_\lambda E_\lambda{}^{(0)} + \sum_\alpha \left[\hbar\omega_{\alpha\lambda}(b_{\alpha\lambda}^+ b_{\alpha\lambda} + \tfrac{1}{2}) \right. \right.$$

$$\left. \left. + C_\lambda^+ C_\lambda g_{\alpha\lambda} \hbar\omega_{\alpha\lambda}(b_{\alpha\lambda} + b_{\alpha\lambda}^+) \right] \right\}$$

and

$$H_T = \langle \lambda' | \tilde{T} | \lambda \rangle + \text{complex conjugate} \qquad\qquad 15.$$

The C_λ^+ and $b_{\alpha\lambda}^+$ are creation operators for fermions and bosons, respectively. $E_\lambda{}^{(0)}$ and $\hbar\omega_{\alpha\lambda}$ are the fermion and boson energies. The coupling is assumed to be linear with a coupling strength $g_{\alpha\lambda}$.

The term H_0 describes the fermion states coupled to their respective boson fields [whose quanta are termed *collectons* (64)]. If the transition is electric dipole then $\tilde{T}$ is the E1 operator.

Duke & Mahan (65) diagonalize H_0 and obtain the following eigenvalues and eigenvectors:

$$E_\lambda = E_\lambda{}^{(0)} - \Delta_\lambda + \sum \hbar\omega_{\alpha\lambda}(n_{\alpha\lambda} + \tfrac{1}{2})$$

where

$$\Delta_\lambda = \sum_\alpha \left[|V_{\alpha\lambda}|^2/\hbar\omega_{\alpha\lambda} \right] \qquad 16.$$

and

$$|\Psi_\lambda\rangle = \exp(S_\lambda)C_\lambda{}^+ |\{n_{\alpha\lambda}\}\rangle$$

where

$$S_\lambda = \sum_\alpha \left\{ [V_{\alpha\lambda}b_{\alpha\lambda} - \overline{V}_{\alpha\lambda}b_{\alpha\lambda}{}^+]/\hbar\omega_{\alpha\lambda} \right\}$$

$$|\{n_{\alpha\lambda}\}\rangle = \prod_\alpha \left[(b_{\alpha\lambda})^{n_{\alpha\lambda}}/(n_{\alpha\lambda}!)^{1/2} \right] |0\rangle$$

and

$$g_{\alpha\lambda}\hbar\omega_{\alpha\lambda} = V_{\alpha\lambda} \qquad 17.$$

(The energy shift Δ_λ due to the coupling between the fermion state and the collective motion is analogous to the *polaron* shift in solids.)

The factor S_λ describes the *collecton correlation* in the wavefunction, which causes the total transition strength associated with a pure transition to spread out into a band of collectons plus single-particle states. This term is directly responsible for the broadening of the transition width. The measure of this broadening is the density of states at the temperature $\theta = (\beta k)^{-1}$, given by:

$$\rho_\beta{}^{(-)}(\lambda, E) = (2\pi\hbar)^{-1} \int_{-\infty}^{\infty} \exp\left\{ i\hbar^{-1}[E + E(\lambda)]t \right\} g_{\beta\lambda}(t)\,dt$$

where

$$E(\lambda) = E_\lambda{}^{(0)} - \Delta_\lambda$$

$$g_{\beta\lambda}(t) = \exp\left\{ -\sum_\alpha |V_{\alpha\lambda}/\hbar\omega_{\alpha\lambda}|^2([\langle n_{\alpha\lambda}\rangle + 1] \right.$$

$$\left. \times [1 - \exp(-i\omega_{\alpha\lambda}t)] + \langle n_{\alpha\lambda}\rangle[1 - \exp(i\omega_{\alpha\lambda}t)]) \right\}$$

and

$$\langle n_{\alpha\lambda}\rangle = [\exp(\beta\hbar\omega_{\alpha\lambda}) - 1]^{-1} \qquad 18.$$

In the special case of the coupling to a single boson field with finite frequency,

$\omega_{\alpha\lambda}$ is replaced by ω_λ and the coupling constant becomes $V_\lambda = g_\lambda \hbar \omega_\lambda$. In this case, a line spectrum for $\rho_\beta{}^{(-)}(\lambda, E)$ occurs:

$$\rho_\beta{}^{(-)}(\lambda, E) = \exp\left[-g_\lambda{}^2(2n_\lambda + 1)\right] \sum_{d=-\alpha}^{\alpha} \delta\left[E + E(\lambda) - d\hbar\omega_\lambda\right]$$
$$\times ([n_\lambda + 1]/n_\lambda)^{d/2} I_d\left\{2g_\lambda{}^2(n_\lambda[n_\lambda + 1])^{1/2}\right\}$$

where

$$n_\lambda = \left[\exp(\beta\hbar\omega_\lambda) - 1\right]^{-1} \qquad\qquad 19.$$

$I_d(x)$ are the modified Bessel functions. The following normalization is required:

$$\int_{-\infty}^{\infty} \rho_\beta{}^{(-)}(\lambda, E)dE = 1 \qquad\qquad 20.$$

In this model, the characteristic broadening of the lineshape exhibits an asymmetry that is most pronounced on the low-energy side of the spectrum (see Ref. 64).

Resonances or fluctuations?—The interpretation of cross-section data obtained in nuclear-reaction studies at high excitation energies is frequently uncertain because of difficulties in distinguishing between resonances and statistical fluctuations. This is particularly true in studies of the giant dipole states in nuclei with $20 \stackrel{<}{\sim} A \stackrel{<}{\sim} 40$ where reasonable estimates of $\langle\Gamma\rangle/\langle D\rangle$, the average total width to average spacing, range from less than unity to greater than ten (66). This corresponds to the transition region between resonances and fluctuations.

Three approaches may help to resolve the problem. In the first, an attempt is made to excite the same nuclear states by different reaction mechanisms. If the energies of the states so produced are correlated, then it is highly probable that the cross section is composed of resonances and not fluctuations due to the random superposition of many overlapping states (67). This method is used by Wu, Firk & Phillips (68) in which their (γ, n_0) results are compared with (p, γ_0) and (γ, p_0) results.

In the second approach, pseudo-cross sections may be computer-generated by using a multilevel theory of nuclear reactions containing suitably chosen parameters (69). The pseudo-cross sections are then compared with the observed cross sections in an attempt to determine the value of $\langle\Gamma\rangle/\langle D\rangle$ that results in the best agreement with the average properties of the observations over wide ranges of energy. Although this may appear to be an indirect approach, it provides one of the only ways of obtaining information on the elusive quantity $\langle\Gamma\rangle/\langle D\rangle$ that does not rely upon dubious extrapolations of the level spacing from much lower energies.

In the third approach, an autocorrelation analysis of the observed cross section is carried out (66). This yields the amplitudes of the fluctuations about the mean and also the value of the mean spacing of the fluctuations. From such information the value of $\langle \Gamma \rangle$ can be obtained (70). A classic example of this approach is afforded by the work of Singh et al (66) in analyzing their results on the ^{27}Al(p,γ_0)^{28}Si reaction.

The following treatment gives the main features of the second approach. The statistical properties of states in the region of fine structure may be investigated by comparing the average properties of the cross sections with those of pseudo-cross sections derived from suitable nuclear-reaction theories (71).

The reduced R-function theory of Thomas (72) provides a particularly useful method of dealing with the ground-state photonuclear cross sections since attention is readily focused on the few reaction channels of interest. The remaining channels may be "eliminated" in the sense of Teichman & Wigner (73).

The (γ,n_0) cross section is proportional to the scattering matrix element $W_{\gamma n_0}$ where:

$$W_{\gamma n_0}(E) \simeq \frac{2iP_\gamma^{1/2}R_{\gamma n_0}P_{\gamma n_0}^{1/2}}{(1 - L_{n_0}R_{n_0 n_0})(1 - L_\gamma R_{\gamma\gamma}) - L_{n_0}L_\gamma R_{\gamma n_0}^2} \qquad 21.$$

in which

$L_{n_0} = S_{n_0} + iP_{n_0}$, the level shift function for neutrons: P_{n_0} is the neutron penetration factor

$L_\gamma = iP_\gamma$, the level shift function for photons: P_γ is the photon penetration factor and the Thomas reduced R function is:

$$R_{\gamma n_0}(E) = \sum_\lambda \frac{\gamma_{\lambda\gamma}\gamma_{\lambda n_0}}{E_\lambda - E - (i/2)\Gamma_\lambda^e} \qquad 22.$$

where

$\gamma_{\lambda\gamma}$ is the incident photon width amplitude

$\gamma_{\lambda n_0}$ is the ground-state neutron width amplitude

Γ_λ^e is the total width of the eliminated channels and

E_λ is the "resonance" energy.

The summation is over all levels λ of the same spin and parity.

If the incident photons are E1 and the outgoing neutrons are s-wave then:

$L_{n_0} = ika$ where k is the neutron wavenumber and a is the effective nuclear radius

$$L_\gamma \propto iE_\gamma^3$$

and

$\Gamma_{\lambda n_0} = 2ka\gamma_{\lambda n_0}^2$ (the ground-state neutron width)

Thomas has shown that Equation 22 is valid even if $\langle \Gamma \rangle / \langle D \rangle \gg 1$ provided

56   FIRK

that the partial widths of the eliminated channels are each smaller than $\langle D \rangle$ and that their amplitudes are random in sign.

With the advent of large, high-speed computers it is now practicable to calculate theoretical cross sections ($\propto |W_{\gamma n_0}|^2$) for many overlapping resonances. The dependence of these cross sections on the values of the input resonance parameters may then be obtained. The quantity $|W_{\gamma n_0}|^2$ is calculated for continuously varying values of $\langle \Gamma \rangle / \langle D \rangle$ typical of those to be expected in the reaction $^{28}Si(\gamma, n_0)^{27}Si$ (68). The values of $\gamma_{\lambda \gamma}$ and $\gamma_{\lambda n_0}$ are randomly sampled from Gaussian distributions with zero mean [this being equivalent to Porter-Thomas distributions in the widths (74)]. The level spacings are chosen from a Wigner-Mehta distribution (75).

The wide energy range covered by the giant dipole resonance (at least 5 MeV) makes it necessary to take into account the variation of the level spacing with energy. This is achieved by using a level-density prescription taken from the work of Halbert et al. (76). The results from this method (68) are found to be in good agreement with those obtained by Singh et al (66) using the third approach.

EXPERIMENTAL DEVELOPMENTS
DIFFERENTIAL CROSS-SECTION AND POLARIZATION MEASUREMENTS

Information concerning the nuclear-structure aspects of the states involved in the reaction can only be obtained by carrying out a number of sophisticated experiments. Among these are measurements of the angular distributions and polarizations of photoparticles and measurements of the γ-ray spectra resulting from the decay of excited states of the residual nuclei. The most important experimental developments in recent years have been made in such measurements. Before discussing the results and their significance, it is necessary to present the general expressions for the differential cross sections and polarizations of photon-induced reactions. These expressions have frequently been limited to those in which only single entrance or exit channels were considered. Early experiments did not warrant more detailed analyses: present experiments, however, require the inclusion of interference effects between different channels (in both initial and final states).

Differential cross sections and polarizations are closely related and will therefore be treated on an equal footing.

General formulas.—Since the early work of Morita et al (77), several detailed presentations of the theory of nuclear reactions involving photons have been given (e.g., Lane & Thomas 71, Baldin et al 78, Welton 79). The present treatment will follow that of (78) with a few minor changes in notation. The reaction of interest is:

$$\gamma + A \rightarrow A^* \rightarrow n + (A - 1)$$

and the quantum numbers involved are shown below:

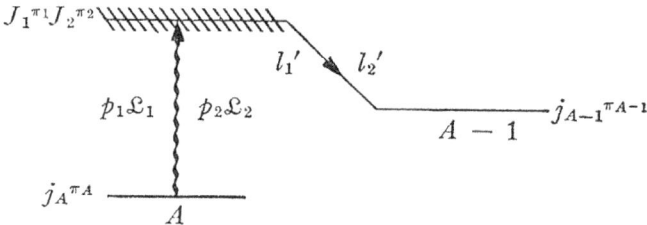

in which:

$j_A{}^{\pi_A}$ are the spin and parity of the target nucleus

$p_1\mathcal{L}_1$ define an incident photon channel (1) leading to an intermediate state of total spin and parity $J_1{}^{\pi_1}$

$p_2\mathcal{L}_2$ define an incident photon channel (2) leading to an intermediate state of total spin and parity $J_2{}^{\pi_2}$

p $=0$ or 1 for magnetic or electric transitions, respectively

$l_1'l_2'$ are the orbital angular momenta of the nucleons decaying from the J_1 and J_2 components, respectively

$j_1'j_2'$ are the intrinsic spins of the nucleons, and

$j_{A-1}{}^{\pi_{A-1}}$ are the spin and parity of the residual nucleus

Additional quantities that will be required are:

$s_1's_2'$ the channel spins associated with l_1' and l_2' respectively, where

$$\mathbf{s_1'} = \mathbf{j_1'} + \mathbf{j_{A-1}} \quad \text{and} \quad \mathbf{s_2'} = \mathbf{j_2'} + \mathbf{j_{A-1}}$$

$$(\text{Note} \quad \mathbf{J_1} = \mathbf{s_1'} + \mathbf{l_1'} \quad \text{and} \quad \mathbf{J_2} = \mathbf{s_2'} + \mathbf{l_2'})$$

and L, the total orbital angular momentum, where

$$\mathbf{L} = \mathbf{l_1'} + \mathbf{l_2'}$$

The differential cross section is then:

$$\frac{d\sigma}{d\Omega} = \frac{\lambda_\gamma^2}{2(2j_A + 1)} \sum \frac{(-1)^{s'-j_{A-1}}}{4} (-1)^{p_1+p_2}(i)^{-l_2'+l_1'-L}$$

$$\times \text{Re}\left\{\langle l_1's'\alpha' | R^{J_1{}^{\pi_1}} | p_1\mathcal{L}_1\alpha \rangle^* \langle l_2's'\alpha | R^{J_2{}^{\pi_2}} | p_2\mathcal{L}_2\alpha \rangle\right\}$$

$$\times Z(l_1'J_1l_2'J_2; s'L)Z_\gamma(\mathcal{L}_1J_1\mathcal{L}_2J_2; j_AL)P_L(\cos\theta)$$

23.

where the sum is over s', p_1, p_2, l_1', l_2', $\mathcal{L}_1$, $\mathcal{L}_2$, J_1, J_2, and L. The vector coupling coefficients Z and Z_γ are defined in (78). Note that in the differential cross section there is no interference between different channel spins, i.e: $s_1' = s_2' = s'$. $P_L(\cos\theta)$ is the usual Legendre polynomial of order L. The indices α and α' in the elements of the reaction matrix $\langle \cdots \alpha' | R^{J^\pi} | \cdots \alpha \rangle$ specify the internal structure of the initial and final states, respectively.

The expression for the differential polarization is more complicated than that given in Equation 23 because the different channel spins s_1' and s_2' must now be explicitly included. We find:

$$\frac{d\mathbf{P}}{d\Omega} = \hat{\mathbf{k}}\,\frac{\lambda_\gamma^2}{4}\,\frac{[2j_1'(j_1'+1)(2j_1'+1)]^{1/2}}{2(2j_A+1)}$$

$$\times \sum \mathrm{Re}\left\{ i\langle l_1's_1'\alpha'\,|\,R^{J_1\pi_1}\,|\,p_1\mathcal{L}_1\alpha\rangle^*\langle l_2's_2'\alpha'\,|\,R^{J_2\pi_2}\,|\,p_2\mathcal{L}_2\alpha\rangle \right\}$$

$$\times (-1)^{j_2'-j_1'+j_A+J_2+s_2'+l_2'-1}Z_\gamma(\mathcal{L}_1 J_1\mathcal{L}_2 J_2;\,j_A L) \qquad\qquad 24.$$

$$\times (-1)^{p_1+p_2}W(j_1's_1'j_1's_2';\,j_2'1)$$

$$\times \left[(2J_1+1)(2l_1'+1)(2s_1'+1)(2J_2+1)(2l_2'+1)(2s_2'+1)\right]^{1/2}$$

$$\times (l_1'0l_2'0\,|\,L0)X(J_1l_1's_1';\,J_2l_2's_2';\,LL1)\overline{P}_L^1(\cos\theta)$$

The summation is over $J_1J_2\mathcal{L}_1\mathcal{L}_2 p_1p_2l_1'l_2's_1's_2'$ and L. The W and X vector coupling coefficients are defined and tabulated in (78). The only nonzero terms in the sum are those for which $p_1+p_2+\mathcal{L}_1+\mathcal{L}_2-L$ is even. $\overline{P}_L^1(\cos\theta)$ is the normalized associated Legendre function (78) and $\hat{\mathbf{k}}$ is a unit vector normal to the scattering plane.

Use of the general formulas.—To illustrate the application of the above formulas, the analysis of data obtained for the $^{16}O(\gamma,n_0)^{15}O$ reaction will be considered. This example is chosen for several reasons: 1. the small number of alternative reaction channels; 2. the availability of reliable differential cross section and polarization measurements, and 3. the availability of theoretical predictions of these quantities which can therefore be tested.

We begin with the assumption of pure E1 absorption. The emitted neutrons must have orbital angular momenta $l=l_1'=l_2'=0$ or 2 since the parities of the intermediate and final states are both negative. The total angular momentum of the intermediate state $J(=J_1=J_2=1)$ is equal to the vector sum of the channel spin $s'(=s_1'=s_2')$ and the orbital angular momenta of the neutrons $l'(=l_1'=l_2'=0$ or 2) so that only channel spin $s'=1$ is allowed. (The value of $s'=1$ also limits the d-wave neutron emission of the $d_{3/2}$ component.) Finally, the total orbital angular momentum $L(=l_1'+l_2')$ is restricted to values $l\leq 2J\leq 2l'$ and also to even values (because the interfering neutron waves have the same parity), hence $L=0$ or 2. The information required to evaluate Equations 23 and 24 is now complete.

In this example, the only matrix elements that contribute are:

$$\langle 01\alpha'\,|\,R^{1-}\,|\,11\alpha\rangle = a_s\exp(i\delta_s)$$

and

$$\langle 21\alpha'\,|\,R^{1-}\,|\,11\alpha\rangle = a_d\exp(i\delta_d),\quad \text{say}$$

where a_s and a_d are the (real) amplitudes for s- and d-wave neutron emission and δ_s and δ_d are their phases, respectively. On evaluating the vector coupling coefficients and carrying out the summations the following expressions are obtained:

$$\frac{d\sigma}{d\Omega} = \frac{3\lambdabar_\gamma{}^2}{16} \left\{ 2(a_s{}^2 + a_d{}^2) + [2 \cdot 2^{1/2} a_s a_d \cos \Delta_{ds} - a_d{}^2] P_2(\cos \theta) \right\} \qquad 25.$$

and

$$\frac{d\mathbf{P}}{d\Omega} = \hat{\mathbf{k}} \lambdabar_\gamma{}^2 0.207 a_s a_d \sin \Delta_{ds} \overline{P}_2{}^1(\cos \theta) \qquad\qquad 26.$$

where $\Delta_{ds} = \delta_d - \delta_s$ is the phase difference.

It is seen that even in this straightforward case there are more unknowns than equations.[2] The best that can be presently achieved is therefore a determination of the ratio a_s/a_d and of the phase difference Δ_{ds}. Nonetheless, a knowledge of these two quantities is an essential beginning for any test of nuclear-structure aspects of the dipole states. Before the results of recent differential cross-section and polarization measurements are presented, a discussion of the methods used to investigate the purity of the E1 absorption process is required. The general forms of the differential cross section and polarization given in Equations 23 and 24 provide a suitable framework within which to work.

Consider the situation in which both E1 and E2 absorption occur, thus leading to a mixed intermediate state. The quantum numbers that arise because of the addition of an E2 component, $|p_2 \mathcal{L}_2 \alpha\rangle = |12\alpha\rangle$, are:

$$J_2{}^{\pi_2} = 2^+, \qquad l_2' = 1 \text{ or } 3, \qquad s_2' = 1, \quad \text{and} \quad L = 0, 1, 2, 3, \text{ and } 4$$

The additional matrix elements that appear are:

$$\langle 11\alpha' | R^{2^+} | 12\alpha \rangle = a_p \exp(i\delta_p)$$

and

$$\langle 31\alpha' | R^{2^+} | 12\alpha \rangle = a_f \exp(i\delta_f), \quad \text{say}$$

where a_p and a_f are the (real) amplitudes for p- and f-wave neutron emission

---

[2] A third linear combination of the unknowns would be obtained by studying the azimuthal distribution of photoparticles which result from the absorption of polarized photons. The angular dependence of the cross section is then found to be:

$$W(\theta, \phi) = 1 + K \sin^2 \theta (1 + P_\gamma \cos 2\phi)$$

where

$$K = \frac{3 \cdot 2^{1/2} |R_d|^2 - 12 \operatorname{Re}(R_d{}^* R_s)}{4 \cdot 2^{1/2} |R_s|^2 + 8 \operatorname{Re}(R_d{}^* R_s) - 2 \cdot 2^{1/2} |R_d|^2}$$

P_γ is the polarization of the incident photons and R_s and R_d are elements of the reaction matrix for s- and d-wave particle emission, respectively.

and δ_p and δ_f are their phases, respectively. The differential cross section now becomes:

$$\frac{d\sigma}{d\Omega} = \frac{\lambdabar_\gamma^2}{8} \sum_{L=0}^{4} A_L P_L(\cos\theta) \qquad 27.$$

where

$A_0 = 3(a_s^2 + a_d^2) + 5(a_p^2 + a_f^2)$

$A_1 = 9.48 a_s a_p \cos\Delta_{ps} - 1.34 a_d a_p \cos\Delta_{pd} + 9.86 a_d a_f \cos\Delta_{fd}$

$A_2 = 2.5 a_p^2 - 1.5 a_d^2 + 2.86 a_f^2 + 3\cdot 2^{1/2} a_s a_d \cos\Delta_{ds} - 1.74 a_p a_f \cos\Delta_{fp}$

$A_3 = 7.74 a_s a_p \cos\Delta_{ps} + 8.04 a_d a_p \cos\Delta_{pd} - 4.38 a_d a_f \cos\Delta_{fd}$

and

$$A_4 = 14 a_p a_f \cos\Delta_{fp} - 2.86 a_f^2$$

in which

$$\Delta_{ps} = \delta_p - \delta_s \quad \text{etc}$$

The coefficients A_L reveal important aspects of the reaction, e.g: A_1 and A_3 contain only E1-E2 interference terms, A_4 contains only pure E2 terms, and A_2 contains only separate E1 and E2 terms. The traditional method of determining the E2 admixture therefore involves a measurement of $d\sigma/d\Omega$ as a function of θ. To obtain significant values of A_L it is necessary to make measurements at more than five angles: this has been done in several cases (80–87).

If an M1 admixture in the absorption process is also allowed, then the problem becomes even more unwieldy. The coefficients A_1 and A_2 then contain additional terms due to M1 effects. However, A_3 and A_4 remain unaffected so that information on the E2 admixture can still be obtained.

An alternative method of detecting the presence of an E2 (or M1 etc) admixture in the absorption process has recently become feasible: the measurement of $d\mathbf{P}/d\Omega$ at appropriate angles θ (88–90). From Equation 26 it is seen that, for pure E1 absorption, $d\mathbf{P}/d\Omega = 0$ at $\theta = 90°$ $(\overline{P}_2^1(\cos\theta) = -(15/16)^{1/2}\sin 2\theta)$: the observation of any polarization at 90° is therefore evidence of the intrusion of other multipoles.

In the case considered above, of mixed E1-E2 absorption, the differential polarization obtained from Equation 24 is found to be:

$$\frac{d\mathbf{P}}{d\Omega} = \hat{\mathbf{k}}\lambdabar_\gamma^2 0.44 \sum_{L=1}^{4} B_L \overline{P}_L^1(\cos\theta) \qquad 28.$$

where

$$B_1 = - 0.79a_s a_p \sin \Delta_{ps} + 0.45a_d a_p \sin \Delta_{pd} + 0.82a_d s_f \sin \Delta_{fd}$$

$$B_2 = 0.47a_s a_d \sin \Delta_{ds} - 0.33a_p a_f \sin \Delta_{fp}$$

$$B_3 = 0.69a_s a_f \sin \Delta_{fs} - 0.72a_d a_p \sin \Delta_{pd} - 0.10a_d a_f \sin \Delta_{fd}$$

$$B_4 = 1.06a_p a_f \sin \Delta_{fp}$$

in which

$$\Delta_{ps} = \delta_p - \delta_s \quad \text{etc}$$

(The absence of a B_0 coefficient reflects the fact that the polarization is a consequence of interference effects.) The previous comments concerning the interpretation of the $A_1 \cdots A_4$ coefficients also apply to the coefficients $B_1 \cdots B_4$.

Examples in ^{16}O.—Precise measurements of the angular distributions of photoprotons from the reaction $^{16}O(\gamma,p_0)^{15}N$ have recently been reported in the energy range between 20 and 30 MeV (84, 86). These measurements were made using silicon solid-state proton spectrometers (at seven angles) with resolutions of approximately 50 keV. In the measurements of Baglin & Thompson (86), the use of a high current, low-duty-cycle linear electron accelerator presented considerable problems due to the high instantaneous counting rates in the solid-state detectors. The steps taken to overcome the difficulties are discussed in (86).

Baglin & Thompson (86) determined the absolute ground-state photoproton cross section in ^{16}O to an accuracy of $\pm 5\%$. The cross section was also measured relative to the $D(\gamma,p)n$ cross section that is known to $\pm 7\%$ in the range of interest (87).

The results of fitting the observed differential cross sections to the expression given in Equation 27 are shown in Figure 6. Baglin & Thompson (86) conclude from these results that the main dipole states decay predominantly by d-wave proton emission following E1 absorption. There is evidence of E2/M1 contributions from the appreciable value of the A_1 coefficient.

Frederick et al (84) obtain a finite value for the A_4 coefficient throughout the entire range, which is to be expected if there is a significant E2 contribution. However, we have already seen that polarization measurements are necessary before more definite conclusions can be drawn. Such measurements have been reported by Hanser (88) who observed the polarization of photoneutrons from ^{16}O at 45° and 90°. This pioneering experiment had two main disadvantages: low resolution which resulted from the use of a helium gas recoil spectrometer to measure the energies; and high backgrounds due to neutron scattering from materials other than the helium gas. More recently, Cole (90) has measured the photoneutron polarization from ^{16}O at 45° by observing the left–right scattering from a liquid helium (scintillation) polarimeter. The neutron energies were measured with good resolution using a nanosecond time-of-flight method. The scattered neutrons

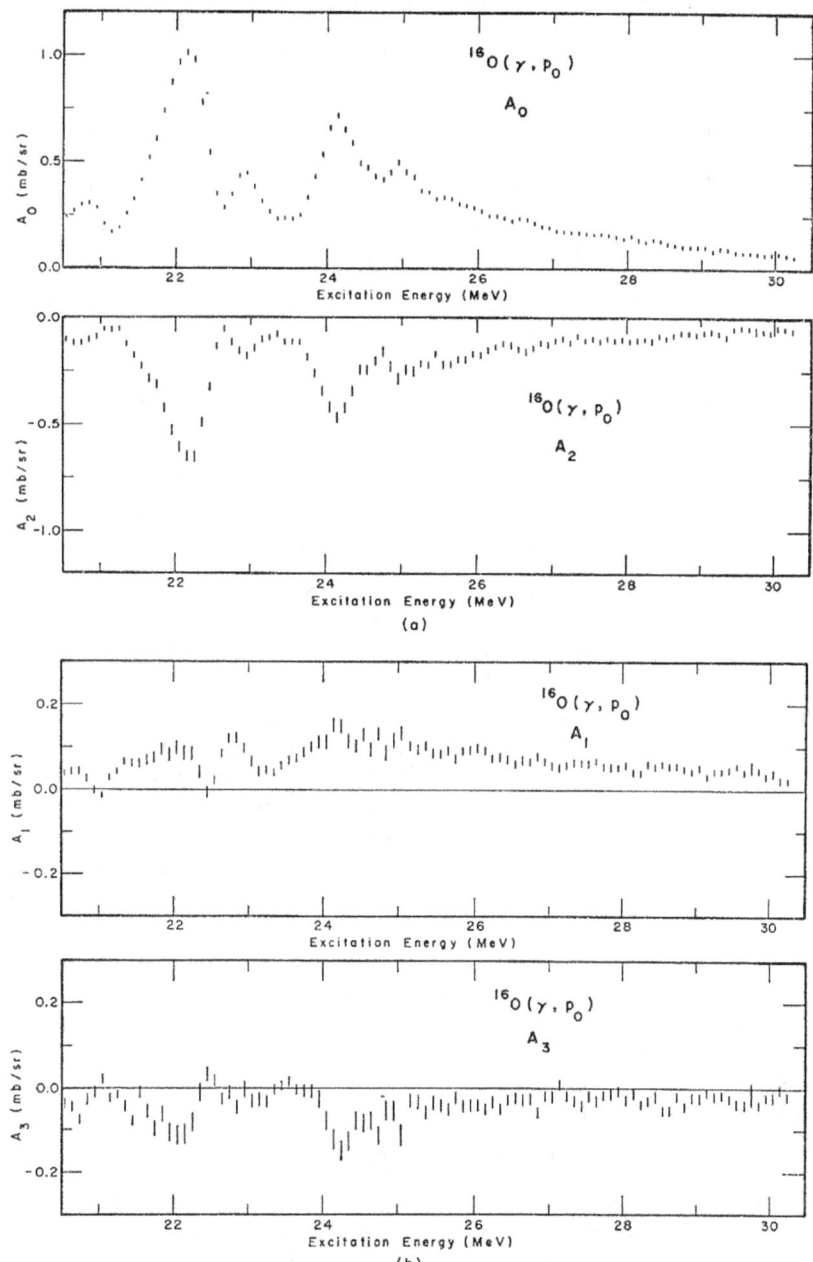

FIGURE 6. (a) Values of the coefficients A_0 and A_2 deduced by fitting the observed differential cross sections for the reaction $^{16}O(\gamma,p_0)^{15}N$ (Baglin & Thompson 86).

(b) Values of the coefficients A_1 and A_3 for the same reaction (86). The value of A_1 quoted in (86) is essentially zero. In a similar experiment, Frederick et al (84) obtain a significant value of A_4, which is clear evidence of E2 absorption.

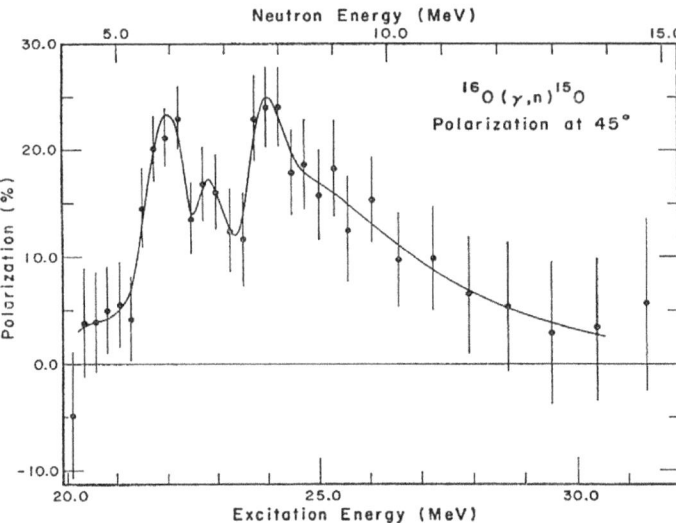

FIGURE 7. The observed polarization from the $^{16}O(\gamma,n)^{15}O$ reaction in the region of the main dipole states determined at a reaction angle of 45° (Cole 90). The line drawn through the points clearly shows the resonant behavior of the polarization.

were detected in coincidence with the helium recoils. The combination of the coincidence method with a nanosecond time-of-flight system resulted in negligible backgrounds. The 45° polarization from $^{16}O(\gamma,n_0)^{15}O$ reaction is shown in Figure 7: the energy dependence follows the known resonant structure (see Figure 6). Assuming, initially, no E2/M1 absorption, the data in Figures 6 and 7 may be analyzed using Equations 25 and 26 to obtain the ratio a_s/a_d and the phase difference Δ_{sd}. In the state at 22.1 MeV, the value of a_s/a_d is then found to be approximately 1/4. Cole (90) and Cole et al (89) also measured the 90° polarization and obtained a value of about -10% thus providing direct evidence of E2/M1 contributions. Unfortunately, not enough independent experiments have been performed, so far, to unravel these finer details of the reaction.

Differential cross sections of photoneutrons from ^{16}O are now being carried out and will add greatly to the reliability of the above analyses (91).

(γ,Nucleon,γ') Reactions

Details of the structure of dipole states may also be obtained by studying the γ-ray spectra that result from the decay of excited states of the residual nuclei populated by nonground-state photonucleon transitions.

Early experiments by Svantesson (92) and Maison et al (93) demonstrated the technique by studying the reactions $^{16}O(\gamma,n\gamma)^{15}O^*$ and $^{16}O(\gamma,p\gamma')^{15}N^*$. The use of low-resolution NaI(Tl) scintillation spectrometers made it difficult to separate many of the closely spaced states in the residual (mirror) nuclei. The development of Ge(Li) γ-ray spectrometers, with their greatly improved resolution and lineshape, removed many of the earlier difficulties. Owens & Baglin (94), Murray & Ritter (95), and Medicus et al (96) have exploited the improved performance of these spectrometers in studies of the γ-ray spectra in ^{15}O, ^{15}N; and ^{11}C, ^{11}B. Quantitative information on the branching ratios of the dipole states to specific residual states has been obtained. The most significant results concern the decay of the 1^- states in ^{16}O and ^{12}C to the positive parity states in the respective residual nuclei [see Figure 8, which clearly shows the positive parity states in ^{15}O and ^{15}N (95)]. These results may be interpreted in terms of $2p$-$2h$ (or higher) configurations present in the dipole states.

An alternative and powerful method has been reported by Caldwell et al (97). They modified the neutron detector, associated with the Livermore photon monochromator, to include NaI(Tl) scintillation spectrometers that could be operated in coincidence (or anticoincidence) with the detected neutrons. In this way, the branching ratios to all the excited states of ^{15}O and ^{15}N up to 10 MeV were obtained. The major fraction of the decays occur to negative parity states of the residual nuclei, which is consistent with an underlying $1p$-$1h$ basis. However, the remaining 15 percent populating positive parity states is now well established and should be taken into account in future descriptions of the dipole states. A recent development that

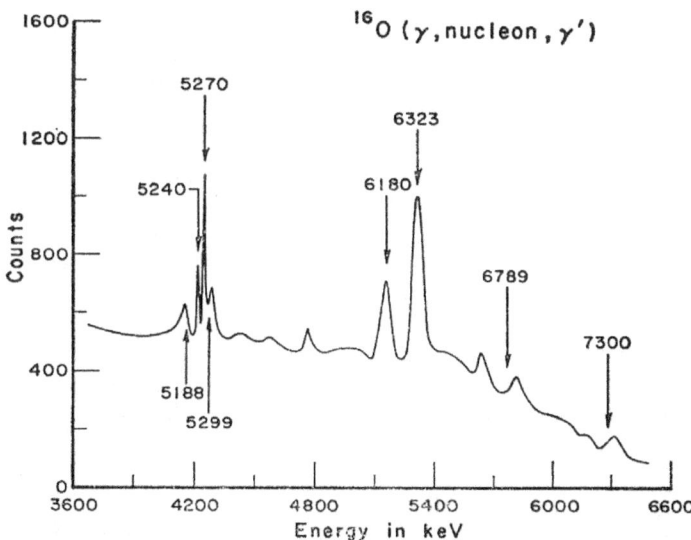

FIGURE 8. The spectra of γ rays from the reactions $^{16}O(\gamma,{}^n_p\gamma')^{15O}_{15N}$ observed using a Ge(Li) detector (95). The transitions from positive parity states in the residual nuclei (at approximately 5200 keV) are due to $2p$-$2h$, or higher, configurations in the dipole states of ^{16}O. The arrows indicate the positions of the double escape peaks in the Ge(Li) detector.

112

should prove informative is the measurement of the angular distribution of γ rays from the excited states of the residual nuclei with respect to the direction of the incident photon beam (98).

THRESHOLD PHOTONEUTRON STUDIES

In nuclei with mass numbers $A \gtrsim 50$, studies of photoneutron cross sections close to the reaction threshold have recently become popular (99–101). The reason for the popularity centers around the high resolution that is attainable using slow neutron time-of-flight spectroscopy (a resolution of $\Delta E \simeq 1$ eV is readily achieved at a neutron energy of 1 keV). Many individual resonances can therefore be resolved and their detailed properties studied. The technique was first used by the MIT electron linear accelerator group about 10 years ago. [In spite of many technical difficulties, primarily associated with a low electron beam intensity, (γ, n_0) cross sections were measured in a number of nuclei (notably the Pb region) and the potentialities of the method thereby clearly demonstrated (102).]

In 1961, Sargent (103) discussed the significance of the resonance parameters obtainable from such measurements. At energies just above threshold, the photoneutrons are predominantly s-wave since the penetrabilities for higher orbital angular momenta are much smaller. Also, the ground-state neutron decay width can generally be separated from other channels in a straightforward manner. The area under an isolated resonance in a (γ, n) cross section curve is proportional to $g\Gamma_n\Gamma_{\gamma0}/\Gamma$ where g is a statistical weighting factor, Γ_n is the neutron width, $\Gamma_{\gamma0}$ is the ground-state radiation width, and Γ is the total width (103). Since the predominant decay mode is by neutron emission, $\Gamma_n/\Gamma \simeq 1$ so that the measured area is directly related to the incident radiation width $\Gamma_{\gamma0}$.

An example of the method is shown in Figure 9 in which the $^{208}\mathrm{Pb}(\gamma, n_0)^{207}\mathrm{Pb}$ cross section is presented (104).

An interesting quantity that can be obtained from these measurements is the strength function $\langle\Gamma_{\gamma0}\rangle/\langle D\rangle$ where $\langle\Gamma_{\gamma0}\rangle$ and $\langle D\rangle$ are the mean values of the ground-state radiation widths and the resonance spacings, respectively. Reliable values of these quantities have been difficult to obtain using more conventional (n, γ_0) measurements. Progress in the (γ, n_0) field is such, however, that significant information in the form of the energy- and A-dependence of $\langle\Gamma_{\gamma0}\rangle/\langle D\rangle$ is now forthcoming.

Other useful pieces of information available from these high-resolution results are the radiation and neutron widths of analog resonances and, in suitable nuclei, precise Coulomb-energy differences (105).

ISOSPIN EFFECTS
SELECTION RULES

It is intuitively clear that comparisons between (γ, n) and (γ, p) reactions should provide information on the charge symmetry of nuclear forces. At low excitation energies, corresponding to the long-wavelength approxima-

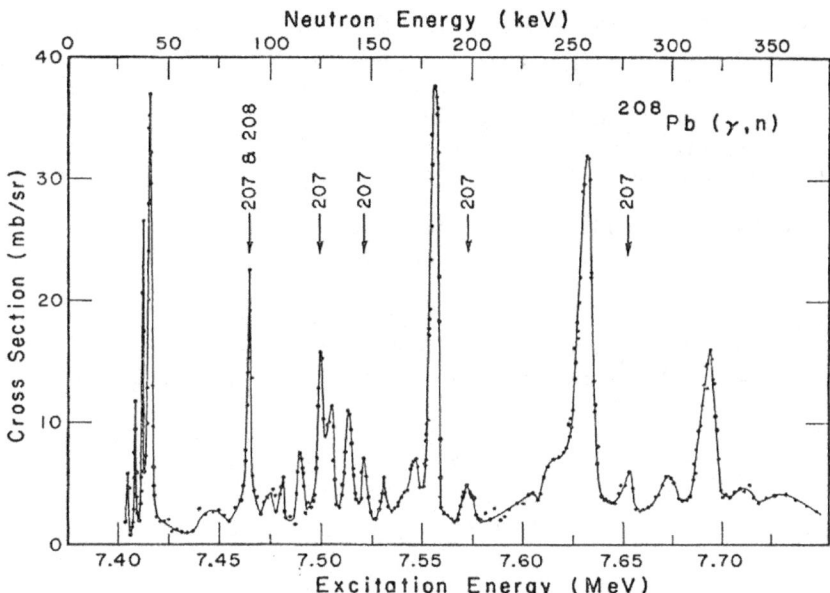

FIGURE 9. The observed cross section for the reaction $^{208}Pb(\gamma,n_0)^{207}Pb$ just above threshold. The asymmetry in the resonance at 41 keV is interpreted in terms of a semidirect process (104).

tion, the properties of the electric dipole operator lead to well-known isospin selection rules. These make the study of photon-induced reactions even more favorable for investigating isospin effects in excited states. Trainor (106) first demonstrated that there is an isospin selection rule for E1 transitions in self-conjugate nuclei $(T = T_z = 0)$:

$$\text{Change in isospin,} \quad \Delta T = 0 \quad \text{No}$$

$$\Delta T = 1 \quad \text{Yes}$$

Radicati (107) and Gell-Mann Telegdi (108) showed that Trainor's result is a special case of more general principles. For example, the non-relativistic Hamiltonian H for the interaction between a photon and a nucleus may be written (neglecting the interaction with magnetic moments):

$$H \simeq (e/2mc) \sum_{k=1}^{A} \mathbf{p}_k \cdot \mathbf{\alpha}(\mathbf{x}_k)(1 - \tau_{zk}) \qquad 29.$$

where p_k is the momentum of the kth nucleon with respect to the center of mass, $\mathbf{\alpha}(\mathbf{x}_k)$ is the vector potential at the position of the kth nucleon, and τ_{zk} is the third component of the isospin operator (with eigenvalues $+1$ for neutrons and -1 for protons).

The Hamiltonian H may be separated into an isoscalar part $H^{(s)}$ (independent of the isospin variables) and an isovector part $H^{(v)}(\tau_{zk})$ thus:

$$H = H^{(s)} + H^{(v)}(\tau_{zk}) \qquad\qquad 30.$$

The general selection rules are then obtained by considering the possible values of the matrix elements between states with initial and final isospins T_i and T_f, respectively, e.g:

$$\langle T_f T_{zf} \mid H^{(v)}(\tau_{zk}) \mid T_i T_{zi}\rangle = (T_i T_{zi}\tau\mu_\tau \mid T_f T_{zf})(2T_f + 1)^{-1/2}$$
$$\times \langle T_f \| H^{(v)}(\tau_{zk}) \| T_i\rangle \qquad\qquad 31.$$

where $H^{(v)}(\tau_{zk})$ transfers isospin τ (components μ_τ) between the initial and final states ($\tau = 1$ for isovectors and $\tau = 0$ for isoscalars).

In the case of E1 transitions in self-conjugate nuclei ($N = Z$, $T_{zi} = T_{zf} = 0$) the selection rules are particularly simple since:

$$H^{(E1)} = (e/2)\sum_k \mathbf{z}_k - (e/2)\sum_k \tau_{zk}\mathbf{z}_k$$
$$= H^{(s)} + H^{(v)} \qquad\qquad 32.$$

and

$$\langle T_f 0 \mid H^{(v)} \mid T_i 0\rangle \propto (T_i 010 \mid T_f 0)$$

The vector coupling coefficient is zero when $T_f = T_i$ and when $T_f \neq T_i \pm 1$ (Trainor's result). These conditions will not be strictly satisfied because of the presence of isospin symmetry-breaking perturbations. The most obvious is due to the Coulomb-energy term:

$$H_{\mathrm{coul}} = (e^2/8)\sum_{k<j}(1 - \tau_{zk})(1 - \tau_{zj})/\mathbf{r}_{kj} \qquad\qquad 33.$$

which does not commute with $\mathbf{T}^2$.

Isospin Mixing in the Dipole States

In 1952, Adair (109) pointed out that the amount of isospin impurity in a given state is generally difficult to determine from studies of the intensities of radiative transitions (radiation widths), because the radiation widths are proportional to $a_{T'}{}^2$ where $a_{T'}$ is the amplitude of the isospin impurity. Typical values of $a_{T'} \simeq 0.1$ are expected for the $T = 0$ amplitudes in the region of the giant dipole ($T = 1$) states of light nuclei (110). Searches for impurities of about 1% in intensity are therefore necessary.

A potentially more sensitive method of measuring $a_{T'}$ was proposed by Barker & Mann (111). It derives from the isospin selection rules discussed above, which state that for E1 photon absorption by a self-conjugate nucleus, only excited states with spin and parity $J^\pi = 1^-$ and $T = 1$ can be formed. However, if the excited state contains a $T = 0$ admixture, the state (formed via its $T = 1$ component) can decay by neutron and proton emission to mirror levels of the residual nuclei via both the $T=0$ and $T=1$ components. Interference effects between them therefore occur so that the emission of neutrons and protons to appropriate final states is sensitive to the amplitudes a_0 and a_1 of the $T=0$ and $T=1$ components, respectively. Small isospin impurities may therefore lead to appreciable differences in the partial photoproton and photoneutron cross sections. Although this feature of photonuclear reactions has been appreciated for some time, only recently have results been reported that are sufficiently precise to take advantage of the method (97, 112, 113).

Barker & Mann (111) studied the problem with particular reference to the ground-state reactions $^{12}C(\gamma,n_0)^{11}C$ and $^{12}C(\gamma,p_0)^{11}B$. By making reasonable assumptions concerning the reaction mechanisms, they deduced a simple relationship between the differential cross-section ratio $d\sigma_p(\theta)$ $/d\sigma_n(\theta)$ and the amplitudes a_0 and a_1, thus:

Let the wavefunction of the excited dipole state in ^{12}C be written:

$$\Psi^* = a_0\Psi_0(\alpha, \ T = 0) + a_1\Psi_1(\alpha, \ T = 1) \qquad 34.$$

where α refers to all other quantum numbers except T. Within a $1p$-$1h$ shell-model basis, the main part of the dipole resonance in ^{12}C is expected to consist, primarily, of the $|1p_{3/2}^{-1}1d_{5/2}\rangle$ configuration. The orbital angular momentum of the outgoing particle is therefore $l=2$ and the channel spin is $s=1$. The reduced width amplitude $\gamma_{t_z s l}$ is then of the form:

$$\gamma_{t_z s l} = \text{const } \delta(s, 1)\delta(l, 2) \sum_{T=0}^{1} a_T(tt_z T^{(A-1)}T_z^{(A-1)} \mid TT_z) \qquad 35.$$

where

$$t_z = +\tfrac{1}{2} \text{ for a neutron and } -\tfrac{1}{2} \text{ for a proton}$$

$$T_z^{(A-1)} = +\tfrac{1}{2} \text{ for } ^{11}B \text{ and } -\tfrac{1}{2} \text{ for } ^{11}C$$

and

$$T_z = 0$$

On evaluating the vector coupling coefficients in Equation 35 the differential cross sections for protons and neutrons become:

$$\frac{d\sigma_p(\theta)}{d\Omega} = \text{const } F_p(\theta)P_p\gamma_p^2 \big| \ 2^{-1/2}(a_1 - a_0) \big|^2 \qquad 36.$$

and

$$\frac{d\sigma_n(\theta)}{d\Omega} = \text{const } F_n(\theta)P_n\gamma_n^2 \big| \ 2^{-1/2}(a_1 + a_0) \big|^2 \qquad 37.$$

where P_p and P_n are the proton and neutron penetrabilities, respectively. Assuming that the proton and neutron distributions have the same angular dependence $(F_p(\theta) = F_n(\theta))$ and that the reduced widths are equal, one obtains the Barker & Mann relationship (with an obvious difference in sign):

$$\frac{d\sigma_p(\theta)}{d\sigma_n(\theta)} = \frac{P_p}{P_n} \left| \frac{a_1 - a_0}{a_1 + a_0} \right|^2 \qquad\qquad 38.$$

For photoparticles well above threshold, the penetration factors approach unity. In such cases, the large differences in the cross-section ratio that result from relatively small values of a_0/a_1 are clearly seen e.g: a ratio of $d\sigma_p(\theta)/d\sigma_n(\theta) \simeq 2/1$ results from a value of $a_0/a_1 \simeq 0.25$ (or 6% in intensity).

Although a number of assumptions are made in deriving Equation 38, it does provide a useful starting point for discussions of possible isospin mixing in the dipole states. (Indeed, it may well give reliable values of the mixing when averaged over suitably wide energy intervals.) The results of recent high-resolution measurements of the 90° differential cross sections for the ground-state reactions $^{12}C(\gamma,n_0)^{11}C$, $^{12}C(\gamma,p_0)^{11}B$, $^{16}O(\gamma,n_0)^{15}O$, $^{16}O(\gamma,p_0)^{15}N$, $^{40}Ca(\gamma,n_0)^{39}Ca$, and $^{40}Ca(\gamma,p_0)^{39}K$ have been reported by Wu, Firk & Phillips (112), Wu et al (113), and Khan et al (114).

In ^{40}Ca, the energy dependence of the two cross sections are remarkably similar. If the data are averaged in intervals as small as 100 keV, the curves appear essentially identical throughout the entire giant resonance. They differ in magnitude, however, by a constant ratio of 2.2:1 in favor of the proton cross section: this corresponds to a value of $a_0/a_1 \simeq 0.25$ (113).

In ^{16}O, the energy dependence of the two cross sections is, again, essentially identical at energies between 21 and 26 MeV (which corresponds to the main giant-resonance region). This time, however, the magnitudes of the peaks differ by less than 15% which implies a high degree of isospin purity $(a_0/a_1 \simeq 0.05)$. Between 19 and 21 MeV appreciable differences between the two cross sections are observed: the cross-section ratio is now strongly energy dependent (even changing sign between the resonances at 19.05 and 19.45 MeV). Unfortunately, the interpretation of these results in terms of isospin mixing is not completely clear-cut because contributions from a number of different multipoles appear at 19 and 20 MeV (115).

In ^{12}C, both the magnitude *and* energy dependence of the two cross sections differ over the main giant-resonance region. The ratio of the proton to neutron cross section reaches a value of 2.5:1 at 22.6 MeV, falling off to a reasonably constant value of 1.5:1 above 26 MeV (112).[3]

---

[3] The 90° differential cross section for the $^{12}C(\gamma,p_0)^{11}B$ deduced from the measurement of Allas et al (81) is 1.5 mb/sr at 22.5 MeV. Brassard (63) has recently questioned the absolute cross-section scale of (81); he suggests that it may be too large by a factor of 1.8. If so, the $^{12}C(\gamma,p_0)^{11}B$ value becomes 0.85 mb/sr in which case the proton/neutron cross section reduces to 1.3:1.

70 FIRK

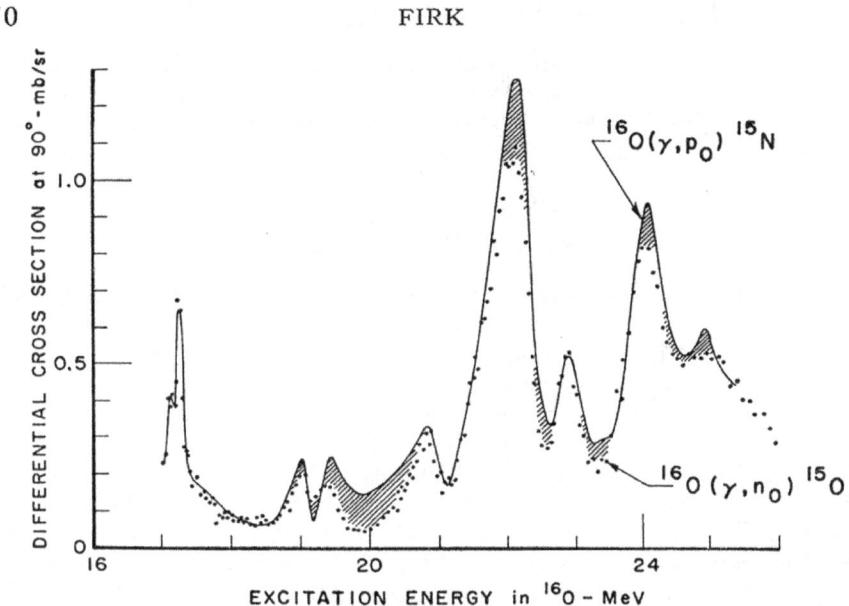

FIGURE 10. A comparison between the 90° differential cross sections for the reactions $^{16}O(\gamma,n_0)^{16}O$ and $^{16}O(\gamma,p_0)^{15}N$ showing the high degree of isospin purity of the main states between 22 and 25 MeV [the (γ,n_0) data are taken from Wu et al (112) and the (γ,p_0) data from Baglin & Thompson (86). The absolute cross sections are known to $\pm 10\%$ in the (γ,n_0) case and to $\pm 5\%$ in the (γ,p_0) case].

The only quantitative predictions of the $(\gamma,p_0)/(\gamma,n_0)$ cross-section ratio have been made in ^{16}O (12, 116). In these two detailed calculations, the average value of the proton cross section is estimated to be more than 25% larger than the neutron cross section. This discrepancy emphasizes the problems of calculating the particle widths of these states.

In an earlier calculation, Greiner (117) explained the asymmetry of the 22.2 MeV state observed in $^{16}O(\gamma,n_0)^{15}O$ and $^{16}O(\gamma,p_0)^{15}N$ reactions in terms of the mixing of $T=0$ and $T=1$ states. The locations and initial shell-model structure of these states were taken from the calculations of Gillet (118) and the particle decay widths were calculated using a prescription based upon R-matrix theory. The proton widths were estimated to be approximately twice the neutron widths, contrary to observations.

ISOSPIN SPLITTING OF THE DIPOLE STATE

In nuclei with $N \neq Z$, the absorption of an E1 photon creates a dipole state with isospin components T and $T+1$ where $T=T_z$ is the isospin of the target nucleus (see Equation 31). This is illustrated in Figure 11.

The energy separation of the two components is given by (119):

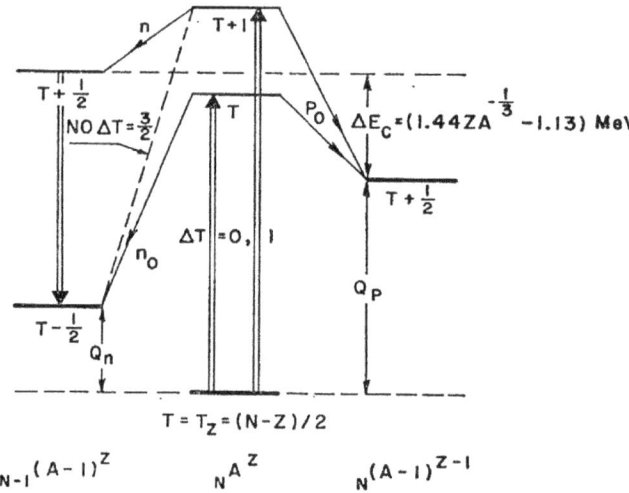

FIGURE 11. Formation and decay of the T and $T+1$ components of the dipole state in nuclei with $T_z \neq 0$.

$$\Delta E_T \simeq \left(\frac{T+1}{T}\right) U \simeq \frac{V}{A}\,(T+1)\ \text{MeV}$$

where U is related to the nuclear-symmetry energy ($U \propto (N-Z)^2/A$) and V is estimated to be 100 MeV (19).

The T component may decay by neutron and proton emission to the ground states of the residual nuclei. However, the $(T+1)$ component can only emit protons to the ground state of the appropriate nucleus because ground-state neutron emission is forbidden. At a certain excitation energy, the first $(T-\frac{1}{2})+1$ state occurs in the residual nucleus $_{N-1}(A-1)^Z$. This state is the analog of the ground state of the residual nucleus $_N(A-1)^{Z-1}$: photoneutrons from the $(T+1)$ component can populate this analog state.

A suitable method of detecting the $(T+1)$ component of the dipole state is therefore to search for the γ-ray transition (or transitions) from the analog state in the residual nucleus as the excitation energy of the target nucleus is raised above the expected location of the $(T+1)$ component.

Murray (120) has recently carried out such an experiment in a search for γ-ray transitions from the $(T+\frac{1}{2})$ states in ^{10}B which would follow from the decay of the $(T+1)$ component of the dipole state in ^{11}B. The particular reaction studied was ^{11}B$(\gamma,n\gamma')^{10}$B*: within the limits set by the sensitivity of the apparatus, no characteristic γ rays were observed, which implied no isospin splitting of the dipole state of ^{11}B.[4]

[4] *Note added in proof:* It has been pointed out that the appropriate states in ^{10}B* are particle unstable so that the reported experiment (120) is not an adequate test of T splitting.

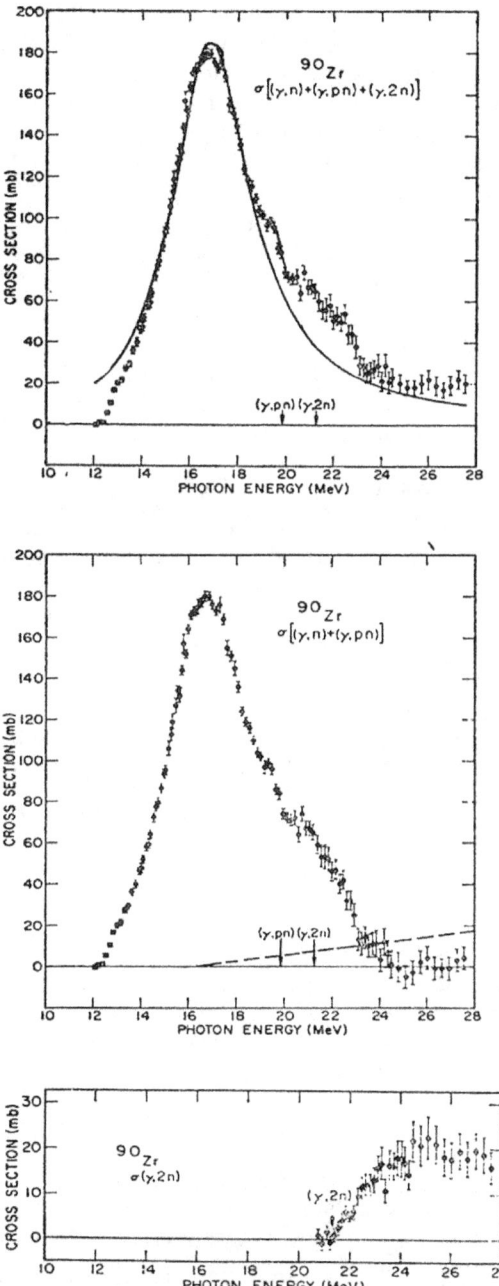

FIGURE 12. The total and partial photoneutron cross sections of ^{90}Zr showing a shoulder on the high-energy side of the giant resonance (24). This has been interpreted as the $T+1$ component of the giant resonance (24, 119).

In ^{90}Zr, the total (γ,n) cross section exhibits a shoulder on the high-energy side of the giant resonance (Berman et al 24) as shown in Figure 12. One of several alternative explanations of the nature of this shoulder, discussed in (24), is that it represents the $(T+1=6)$ component of the dipole state. [Fallieros and Goulard had previously predicted the existence of this component at an excitation energy of approximately 21 MeV having an integrated strength of 20% of the main $(T=5)$ component.] The interpretation of the shoulder in terms of isospin splitting is now much firmer following the observation of a number of sharp analog states in the reaction ^{89}Y$(p,\gamma_0)^{90}$Zr at excitation energies in the region of 21 MeV (121).

Min (122) has suggested that isospin splitting of the dipole state occurs in ^{58}Ni. He associates the increased yield of photoprotons from the region around 21 MeV with the $(T=2)$ component while the main $(T=1)$ component is centered at 17 MeV. More definitive experiments are required before this splitting is established beyond reasonable doubt.

AN ISOSPIN SUM RULE

Goulard & Fallieros (123) have shown that the expression for the relative strengths of the two T components of the dipole state must include terms from nuclear-structure effects in addition to those from the vector coupling coefficients given in Equation 31. For example, certain p-h configurations were shown to effect $\Delta T = 0$ but not $\Delta T = 1$ transitions.

O'Connell (124) has derived a sum rule that includes the Goulard-Fallieros result as a special case. The difference between the bremsstrahlung-weighted cross sections of the T and $(T+1)$ components is shown to be related to the mean square isovector $\langle R_V{}^2\rangle$ and mean square isotensor $\langle R_T{}^2\rangle$ radii as follows:

$$\frac{\pi^2 e^2}{3\hbar c}\left[2T\langle R_V{}^2\rangle + T(2T-1)\langle R_T{}^2\rangle\right] = \sigma_{-1}(T) - T\sigma_{-1}(T+1) \qquad 39.$$

where

$$\sigma_{-1}(T) = \int \sigma(T)\,\frac{dE}{E}$$

$$2T_z\langle R_V{}^2\rangle = N\langle R_n{}^2\rangle - Z\langle R_p{}^2\rangle$$

and

$$[3T_z{}^2 - T(T+1)]\langle R_T{}^2\rangle = \left\langle TT_z \left| \sum_{j<k}(\mathbf{r}_j\cdot\mathbf{r}_k)(3\tau_{zj}\tau_{zk} - \boldsymbol{\tau}_j\cdot\boldsymbol{\tau}_k) \right| TT_z \right\rangle$$

in which $\langle R_n{}^2\rangle$ and $\langle R_p{}^2\rangle$ are the mean square neutron and proton radii, respectively.

The isotensor radius depends on the spatial correlations between pairs of nucleons, j and k. It is also directly related to the correlations between pairs of excess neutrons because the term involving the nucleon isospins yields a value of zero for the $T = 0$ core.

In a pure independent-particle model, there are no correlations between pairs of neutrons, in which case $\langle R_T{}^2 \rangle = 0$. The sum rule then becomes:

$$\frac{\pi^2 e^2}{3\hbar c} 2T \langle R_V{}^2 \rangle = \sigma_{-1}(T) - T\sigma_{-1}(T+1) \qquad 40.$$

so that

$$\frac{\sigma_{-1}(T+1)}{\sigma_{-1}} = \frac{1}{T+1} \left[1 - \left\{ \frac{2\pi^2 e^2 T \langle R_{ch}{}^2 \rangle}{3\hbar c \sigma_{-1}} \right\} \right] \qquad 41.$$

where

$$\sigma_{-1} = \sigma_{-1}(T) + \sigma_{-1}(T+1)$$

and

$$2T_z \langle R_V{}^2 \rangle = 2T \langle R_{ch}{}^2 \rangle$$

It is here assumed that $\langle R_n{}^2 \rangle = \langle R_p{}^2 \rangle$ and that these are equal to the mean square charge radius $\langle R_{ch}{}^2 \rangle$. The bremsstrahlung-weighted cross section for a harmonic oscillator is:

$$\sigma_{-1}{}^{(h.o.)} = \int \sigma \frac{dE}{E} \simeq \left(0.06 \frac{NZ}{A} \times \frac{1}{41 A^{-1/3}} \right) \text{ barn} \qquad 42.$$

On substituting this value in Equation 41 the Goulard-Fallieros expression is obtained.

In practice the factor 41 in the expression for the resonance energy in a harmonic-oscillator model should be almost doubled. This makes the structure term in Equation 41 even more important in reducing the size of the $(T+1)$ component.

ANALOG STATES

The excitation of analog states in the (γ, p) process has been recently demonstrated by Shoda et al (125). In their work, the states are excited by virtual photons using the $(e, e'p)$ reaction. The photoproton spectra are measured with a resolution of about 50 keV using a multichannel magnetic spectrometer. Estimates of the radiative widths are made and compared with the theoretical estimates of Goulard et al (126). In most cases, the agreement is satisfactory.

Studies of the inverse (p, γ) reactions have played an important part in our understanding of analog states. A comprehensive account of these measurements has been given by Hanna (127) and they will therefore be omitted from the present discussion.

ACKNOWLEDGMENT

I am indebted to all members of the Electron Accelerator Laboratory at Yale who have contributed so much to the work reported here. In particular, I wish to thank Drs. John Baglin, Barry Berman, Claude Brassard, George Cole, Jr., and Chung-Pao Wu for providing results and expert advice. Dr. Evans Hayward (National Bureau of Standards, Washington, D.C.) has frequently supplied new information. Finally, it is a pleasure to acknowledge the enthusiastic support given to this work by Professor Howard Schultz.

LITERATURE CITED

1. Heitler, W. 1947. *The Quantum Theory of Radiation*. London: Oxford. 2nd ed.
2. Moszkowski, S. A. 1965. In *Alpha-, Beta- and Gamma-Ray Spectroscopy*, ed. K. Siegbahn, vol. 2. Amsterdam: North-Holland
3. Baldwin, G. C., Klaiber, G. S. 1948. *Phys. Rev.* 73:1156
4. Levinger, J. S. 1960. *Nuclear Photo-Disintegration*. London: Oxford
5. Goldhaber, M., Teller, E. 1948. *Phys. Rev.* 74:1046
6. Fuller, E. G., Hayward, E. 1962. In *Nuclear Reactions*, ed. P. M. Endt, P. B. Smith, II. Amsterdam: North-Holland
7. Spicer, B. M. 1969. In *Advances in Nuclear Phys.*, ed. M. Baranger, E. Vogt, II. New York: Plenum
8. Bohr, A., Mottelson, B. R. *Nuclear Theory*. II. New York: Benjamin. To be published
9. Hayward, E. 1963. *Rev. Mod. Phys.* 35:234
10. Firk, F. W. K. 1964. *Nucl. Instr. Meth.* 28:205
11. Elliott, J. P., Flowers, B. H. 1957. *Proc. Roy. Soc. London A*, 242:57
12. Buck, B., Hill, A. D. 1967. *Nucl. Phys.* A95:271
13. Mahaux, C., Weidenmüller, H. A. 1969. *Shell-Model Approach to Nuclear Reactions*. Amsterdam: North-Holland
14. Strauch, K. 1953. *Ann. Rev. Nucl. Sci.* 2:105
15. Levinger, J. S. 1954. *Ann. Rev. Nucl. Sci.* 4:13
16. Wilkinson, D. H. 1959. *Ann. Rev. Nucl. Sci.* 9:1
17. Danos, M., Fuller, E. G. 1965. *Ann. Rev. Nucl. Sci.* 15:29
18. Bohr, A., Mottelson, B. R. 1969. *Nuclear Theory* I. New York: Benjamin
19. Lane, A. M. 1962. *Nucl. Phys.* 35:676
20. Levinger, J. S., Bethe, H. A. 1950. *Phys. Rev.* 78:115
21. Fultz, S. C., Bramblett, R. L., Caldwell, J. T., Kerr, N. A. 1962. *Phys. Rev.* 127:1273
22. Bramblett, R. L., Caldwell, J. T., Auchampaugh, G. F., Fultz, S. C. 1963. *Phys. Rev.* 129:2723
23. Bramblett, R. L., Caldwell, J. T., Berman, B. L., Harvey, R. R., Fultz, S. C. 1966. *Phys. Rev.* 148:B1198
24. Berman, B. L., Caldwell, J. T., Har-

vey, R. R., Kelly, M. A., Bramblett, R. L., Fultz, S. C. 1967. *Phys. Rev.* 162:1098
25. Berman, B. L., Kelly, M. A., Bramblett, R. L., Caldwell, J. T., Davis, H. S., Fultz, S. C. 1969. *Phys. Rev.* 185:1576
26. Fultz, S. C., Berman, B. L., Caldwell, J. T., Bramblett, R. L., Kelly, M. A. 1969. *Phys. Rev.* 186:1255
27. Tzara, C. 1957. *Compt. Rend.* 245:56
28. Schuhl, C., Tzara, C. 1961. *Nucl. Instr. Meth.* 10:217
29. Jupiter, C. P., Hansen, N. E., Shafer, R. E., Fultz, S. C. 1961. *Phys. Rev.* 121:866
30. Bergère, R., Beil, H., Veyssière, A., 1968. *Nucl. Phys.* A121:463
31. Okamoto, K. 1958. *Phys. Rev.* 110:143
32. Danos, M. 1958. *Nucl. Phys.* 5:23
33. Steinwedel, H., Jensen, J. H. D. 1950. *Z. Naturforsch.* 5:413
34. Fuller, E. G., Weiss, M. S. 1958. *Phys. Rev.* 112:291
35. Spicer, B. M., Thies, H. H., Baglin, J. E. E., Allum, F. R. 1958. *Aust. J. Phys.* 11:298
36. Olesen, M. C., Elbek, B. 1960. *Nucl. Phys.* 15:134
37. Carrigan, R. A., Jr., Gupta, P. D., Sutton, R. B., Suzuki, M. N., Thompson, A. C., Coté, R. E., Prestwich, W. V., Gaigalas, A. K., Raboy, S. 1968. *Phys. Rev. Lett.* 20:874
38. Stelson, P. H., Grodzins, L. 1965. *Nucl. Data* A1:21
39. McGowan, F. K., Stelson, P. H. 1958. *Phys. Rev.* 109:901
40. Bernstein, E. M., Graetzer, R. 1960. *Phys. Rev.* 119:1321
41. Barrett, R. C., Bernow, S., Devons, S., Duerdoth, I., Hitlin, D., Kast, J. W., Lee, W. Y., Macagno, E. R., Rainwater, J., Wu, C. S. *Columbia Univ. Pegram Lab. Rept: NYO 72-191 1968*. Unpublished
42. Stokstad, R. G., Persson, B. 1968. *Phys. Rev.* 170:1072
43. Kelly, M. A., Berman, B. L., Bramblett, R. L., Fultz, S. C. 1969. *Phys. Rev.* 179:1194
44. Ambler, E., Fuller, E. G., Marshak, H. 1965. *Phys. Rev.* 138:117
45. Arenhövel, H., Danos, M., Greiner, W. 1967. *Phys. Rev.* 157:1109
46. Brown, G. E., Bolsterli, M. 1959. *Phys. Rev. Lett.* 3:472
47. Kuo, T. T. S., Blomqvist, J., Brown, G. E. 1970. *Phys. Lett.* 31B:93

48. Kuo, T. T. S., Brown, G. E. 1967. *Nucl. Phys.* A92:481
49. Gillet, V., Green, A. M., Sanderson, E. A. 1966. *Nucl. Phys.* 88:321
50. Blomqvist, J., Kuo, T. T. S. 1969. *Phys. Lett.* 29B: 544
51. Kamimura, M., Ikeda, K., Arima, A. 1967. *Nucl. Phys.* A95:129
52. Rowe, D. J. 1968. *Rev. Mod. Phys.* 40:153
53. Nilsson, S. G., Sawicki, J., Glendenning, N. K. 1962. *Nucl. Phys.* 33:239
54. Drechsel, D., Seaborn, J. B., Greiner, W. 1968. *Phys. Rev. Lett.* 17:488
55. Rowe, D. J., Wong, S. S. M. 1969. *Phys. Lett.* 30B:147
56. Wong, S. S. M., Rowe, D. J. 1969. *Phys. Lett.* 30B:150
57. Saruis, A. M., Marangoni, M. 1969. *Nucl. Phys.* A132:433
58. Beres, W. P., MacDonald, W. M. 1967. *Nucl. Phys.* A91:529
59. Danos, M., Greiner, W. 1966. *Phys. Rev.* 146:708
60. Wahsweiler, W. G., Danos, M., Greiner, W. 1968. *Phys. Rev.* 170:983
61. Greiner, W. Private communication
62. Boeker, E. 1963. Doctoral thesis. Vrije Univ., Amsterdam. Unpublished
63. Brassard, C. Doctoral thesis. Yale Univ., New Haven, Conn. Unpublished
64. Duke, C. B., Malik, F. B., Firk, F. W. K. 1967. *Phys. Rev.* 157:879
65. Duke, C. B., Mahan, G. D. 1965. *Phys. Rev.* 139:A1965
66. Singh, P. P., Segel, R. E., Meyer-Schützmeister, L., Hanna, S. S., Allas, R. G. 1965. *Nucl. Phys.* 65:577
67. Ericson, T. E. O. 1960. *Phys. Rev. Lett.* 5:430
68. Wu, C. P., Firk, F. W. K., Phillips, T. W., *Nucl. Phys.* To be published
69. Lynn, J. E. 1966. In *Nuclear Structure Study with Neutrons*, ed. M. Nève de Mévergnies. Amsterdam: North-Holland
70. Brink, D. M., Stephen, R. O. 1963. *Phys. Lett.* 5:77
71. Lane, A. M., Thomas, R. G. 1958. *Rev. Mod. Phys.* 30:257
72. Thomas, R. G. 1955. *Phys. Rev.* 97:224
73. Teichman, T., Wigner, E. P. 1952. *Phys. Rev.* 87:123
74. Porter, C. E., Thomas, R. G. 1956. *Phys. Rev.* 104:483
75. Mehta, M. L. 1960 *Nucl. Phys.* 18:395
76. Halbert, M. L., Durham, F. E., Moak, C. D., Zucker, A. 1963. *Nucl. Phys.* 47:353
77. Morita, M., Sugie, A., Yoshida, S. 1954. *Progr. Theor. Phys.* 12:713
78. Baldin, A. M., Goldanskii, V. I., Rozenthal, I. L. 1961. *Kinematics of Nuclear Reactions*. London: Oxford
79. Welton, T. A. 1963. In *Fast Neutron Physics*, ed. J. B. Marion, J. L. Fowler, II. New York: Interscience
80. Tanner, N. W., Thomas, G. C., Earle, E. D. 1964. *Nucl. Phys.* 52:29
81. Allas, R. G., Hanna, S. S., Meyer-Schützmeister, L., Segel, R. E. 1964. *Nucl. Phys.* 58:122
82. Dearnaley, G., Gemmell, D. S., Hooton, B. W., Jones, G. A. 1965. *Nucl. Phys.* 64:177
83. Earle, E. D., Tanner, N. W. 1967. *Nucl. Phys.* A95:241
84. Frederick, D. E., Stewart, R. J. J., Morrison, R. C. 1969. *Phys. Rev.* 186:992
85. Frederick, D. E., Sherick, A. D. 1968. *Phys. Rev.* 176:1177
86. Baglin, J. E. E., Thompson, M. N. 1969. *Nucl. Phys.* A138:73
87. Weissman, B., Schultz, H. L. *Phys. Rev. Lett.* To be published
88. Hanser, F. 1967. Doctoral thesis. MIT, Cambridge, Mass.
89. Cole, G. W., Jr., Firk, F. W. K., Phillips, T. W. 1969. *Phys. Lett.* 30B:91
90. Cole, G. W. Jr. 1970. Doctoral thesis. Yale Univ., New Haven, Conn.
91. Jury, J. W. 1970. Doctoral thesis. Univ. Toronto
92. Svantesson, N. L. 1957. *Nucl. Phys.* 3:273
93. Maison, J. M., Langevin, M., Loiseaux, J. M. 1965. *Phys. Lett.* 19:308
94. Owens, R. O., Baglin, J. E. E. 1966. *Phys. Rev. Lett.* 17:524
95. Murray, K. M., Ritter, J. 1969. *Phys. Rev.* 182:1097
96. Medicus, H. A., Bowey, E. M., Gayther, D. B., Patrick, B. H., Winhold, E. J. Private communication
97. Caldwell, J. T., Fultz, S. C., Bramblett, R. L. 1967. *Phys. Rev. Lett.* 19:447
98. McConnell, D. B. Private communication
99. Bowman, C. D., Sidhu, G. S., Berman, B. L. 1967. *Phys. Rev.* 163:951
100. Patrick, B. H. Private communication
101. McNeill, K. G. Private communication
102. Bertozzi, W., Sargent, C. P., Turchinetz, W. 1963. *Phys. Lett.* 6:108

78   FIRK

103. Sargent, C. P. 1961. In *Neutron Time-of-Flight Methods*, ed. J. Spaepen. Brussels: EANEC
104. Bowman, C. D., Baglan, R. J., Berman, B. L. 1969. *Phys. Rev. Lett.* 23:796
105. Berman, B. L., Baglan, R. J., Bowman, C. D. 1970. *Phys. Rev. Lett.* 24:319
106. Trainor, L. E. H. 1952. *Phys. Rev.* 85:962
107. Radicati, L. A. 1952. *Phys. Rev.* 87:521
108. Gell-Mann, M., Telegdi, V. L. 1953. *Phys. Rev.* 91:169
109. Adair, R. K. 1952. *Phys. Rev.* 87:1041
110. MacDonald, W. M. 1956. *Phys. Rev.* 101:271
111. Barker, F. C., Mann, A. K. 1957. *Phil. Mag.* 2:5
112. Wu, C.-P., Firk, F. W. K., Phillips, T. W. 1968. *Phys. Rev. Lett.* 20:1182
113. Wu, C.-P., Baglin, J. E. E., Firk, F. W. K., Phillips, T. W. 1969. *Phys. Lett.* 29B:359
114. Khan, T. A., Hewitt, J. S., McNeill, K. G. 1969. *Can. J. Phys.* 47:1037
115. Goldman, A. Private communication
116. Perez, J. D., MacDonald, W. M. 1969. *Phys. Rev.* 182:1066
117. Greiner, W. 1963. *Nucl. Phys.* 49:522
118. Gillet, V. 1962. Doctoral thesis. Univ. Paris
119. Fallieros, S., Goulard, B., Venter, R. H. 1965. *Phys. Lett.* 19:398
120. Murray, K. M. 1969. *Phys. Rev. Lett.* 23:1461
121. Hasinoff, M., Fisher, G. A., Kuan, H. M., Hanna, S. S. Private communication
122. Min, K. 1969. *Phys. Rev.* 182:1359
123. Goulard, B., Fallieros, S. 1967. *Can. J. Phys.* 45:3221
124. O'Connell, J. S. 1969. *Phys. Rev. Lett.* 22:1314
125. Shoda, K., Sugawara, M., Saito, T., Miyase, H. 1969. *Phys. Rev. Lett.* 23:800
126. Goulard, B., Hughes, T. A., Fallieros, S. 1968. *Phys. Rev.* 176:1345
127. Hanna, S. S. 1969. In *Isospin in Nuclear Physics*, ed. D. H. Wilkinson. Amsterdam: North-Holland

Neutron Polarization

F. W. K. Firk

Electron Accelerator Laboratory, Yale University, New Haven, Ct. 06520

RÉSUMÉ

Some recent experiments involving polarized neutrons are discussed; they demonstrate how polarization studies provide information on fundamental aspects of nuclear structure that cannot be obtained from more traditional neutron studies.

ABSTRACT

Until recently, neutron polarization studies tended to be limited either to very low energies or to restricted regions at higher energies, determined by the kinematics of favorable $(p,\vec{n})$ and $(d,\vec{n})$ reactions. With the advent of high intensity pulsed electron and proton accelerators and of beams of vector polarized deuterons, this is no longer the case. We have entered an era in which neutron polarization experiments are now being carried out, in a routine way, throughout the entire range from thermal energies to tens-of-MeV. The significance of neutron polarization studies is illustrated in discussions of a wide variety of experiments that include i) the measurement of T-invariance in the β-decay of polarized neutrons ii) a search for the effects of meson exchange currents in the photo-disintegration of the deuteron iii) the determination of quantum numbers of states in the fission of aligned ^{235}U and ^{237}Np induced by polarized neutrons and iv) the double- and triple-scattering of fast neutrons by light nuclei.

INTRODUCTION

Studies of polarization effects in nuclear reactions involving neutrons provide information of a basic nature that can be obtained only indirectly (or sometimes not at all) using traditional experimental methods. We recall the work of Adair et al[1] at Wisconsin in the early 50's in which the sign and magnitude of the nuclear spin-orbit potential was first established by studying the polarization of neutrons scattered from various nuclei.

Polarization is a consequence of interference effects between the amplitudes associated with a particular process. In neutron induced reactions, such effects can arise in many ways, for example: i) in non-resonant scattering at those energies where many different partial waves are allowed ii) from interference between certain resonant and non-resonant scattering iii) from resonance - resonance interference iv) from interference from the cumulative effect of distant levels (which may be interpreted using an optical model) v) from the presence of a spin-spin term. In photon induced reactions,[2,3] polarization effects can arise from interference between multipoles of appropriate angular momentum and parity and between photonucleon decay channels (either from different states or from an isolated state which is a superposition of base states of different relative orbital angular momentum). The results of such studies can therefore elucidate fundamental questions of nuclear structure.

Perhaps the best-known examples of the essential part played by polarization studies in Nuclear Physics involve tests of P- and T-invariance of quantum systems [4,5] and of the basic features of the nucleon-nucleon interaction.[6]

The following examples have been chosen to illustrate the wide variety of information that has been obtained in this field lately. (Several detailed reviews of neutron polarization have been given in the past [see Haeberli[7], Barschall[8] and Walter[9]]).

PRINCIPLES OF POLARIZATION

We shall limit the discussion to spin 1/2-spin 0 elastic scattering. Before scattering, the spin-part of the neutron wave function is

$$\chi_1 = \begin{pmatrix} \alpha_1^\uparrow \\ \alpha_1^\downarrow \end{pmatrix}$$

where $\alpha_1^\uparrow$ and $\alpha_1^\downarrow$ are the complex amplitudes for the spin to be "up" or "down", respectively.

After scattering (states which we label with primes), the amplitudes are changed, giving

$$\chi_1' = \begin{pmatrix} \alpha_1^{\uparrow'} \\ \alpha_1^{\downarrow'} \end{pmatrix}$$

where $\quad \tilde{M}_1 \chi_1 = \chi_1'$

If the scattering matrix $\tilde{M}_1$ is to be invariant under rotations and reflections (conservation of angular momentum and of parity) it must have the form[10]

$$\underset{1}{\tilde{M}} = g_1 \underset{1}{\overset{\sim}{1}} + h_1 \vec{\sigma} \cdot \hat{n}_1 \tag{1}$$

where
$$\overset{\sim}{1} = \begin{pmatrix} 1 & 0 \\ 0 & 1 \end{pmatrix} , \vec{\sigma} \text{ is the Pauli spin (vector) matrix}$$

and
$$\hat{n}_1 = \vec{k}_1 \times \vec{k}_1' / |\vec{k}_1 \times \vec{k}_1'| \text{ is a unit vector normal to}$$

the scattering plane. Here, $\vec{k}_1$ and $\vec{k}_1'$ are the momenta before and after the (first) scattering, respectively.

We shall see that the state of polarization after the first scattering can be determined by scattering from a second nucleus of known analyzing power. This process results in yet another change in the amplitudes:

$$\chi_2' = \begin{pmatrix} \alpha_2^{\uparrow'} \\ \alpha_2^{\downarrow'} \end{pmatrix}$$

where
$$\underset{2}{\tilde{M}} \chi_2 = \chi_2'$$

and
$$\chi_2 = \chi_1'$$

The scattering matrix $\underset{2}{\tilde{M}}$ has the form

$$\underset{2}{\tilde{M}} = g_2 \overset{\sim}{1} + h_2 \vec{\sigma} \cdot \hat{n}_2 \tag{2}$$

where
$$\hat{n}_2 = \vec{k}_2 \times \vec{k}_2' / |\vec{k}_2 \times \vec{k}_2'| \text{ is a unit vector normal to the}$$

second scattering plane

and
$$\vec{k}_2 = \vec{k}_1'$$

The two planes are rotated with respect to each other by the angle ϕ, so that

$$\hat{n}_1 \cdot \hat{n}_2 = \cos \phi$$

If the initial beam is unpolarized $(< |\alpha_1^{\uparrow}|^2 - |\alpha_1^{\downarrow}|^2 > = 0$ etc. $)$

the differential cross section after the first scattering is

$$\frac{d\sigma_1}{d\Omega} = |g_1|^2 + |h_1|^2$$

and the state of polarization of the beam after the first scattering is

$$\frac{\left|\alpha_1^{\uparrow'}\right|^2 - \left|\alpha_1^{\downarrow'}\right|^2}{\left|\alpha_1^{\uparrow'}\right|^2 + \left|\alpha_1^{\downarrow'}\right|^2} = \left(\frac{g_1 h_1^* + g_1^* h_1}{\left|g_1\right|^2 + \left|h_1\right|^2}\right) \cos \phi \tag{3}$$

$$= p_1 \cos \phi$$

where
$$\vec{p}_1 = p_1 \hat{n}_1$$

The beam is seen to be polarized if $h_1 \neq 0$.

(The complex amplitudes g_1 and h_1 are called the non-spin-flip and spin-flip amplitudes, respectively. If $h_1 \neq 0$, it means that there is a non-central part to the potential (eg: a spin-orbit part)).

If this partially polarized beam is scattered a second time, the differential cross section is found to be

$$\frac{d\sigma_2}{d\Omega} = (\left|g_1\right|^2 + \left|h_1\right|^2)(1 + \vec{p}_1 \cdot \vec{A}_2) \tag{4}$$

where
$$\vec{A}_2 = A_2 \hat{n}_2 \text{ is the analyzing power of the second scatterer}$$

and
$$A_2 = \frac{g_2 h_2^* + g_2^* h_2}{\left|g_2\right|^2 + \left|h_2\right|^2} \tag{5}$$

therefore
$$\frac{d\sigma_2}{d\Omega} = \left(\frac{d\sigma_2}{d\Omega}\right)_{unpol} (1 + \vec{p}_1 \cdot \vec{A}_2) \tag{6}$$

and
$$\vec{p}_1 \cdot \vec{A}_2 = p_1 A_2 \cos \phi$$

The method of measuring p_1 is therefore to observe the left-right symmetry, in scattering from a second target (2), in the same plane as the first reaction ie: the angle ϕ is 0 or π. In this case

$$\frac{d\sigma_2}{d\Omega} = \left(\frac{d\sigma_2}{d\Omega}\right)_{unpol} (1 \pm p_1 A_2) \tag{7}$$

and
$$\frac{d\sigma^R - d\sigma^L}{d\sigma^R + d\sigma^L} = p_1 A_2 = R \,. \tag{8}$$

The sign convention is as shown [11)]

130

[Looking down on the scattering plane]

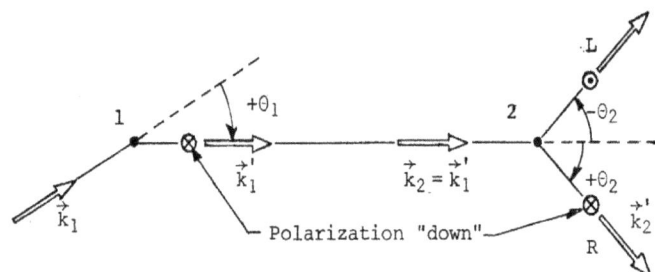

In elastic scattering, the polarization is equal to the analyzing power (p=A).

In order to measure p_1, we require a measurement of R and of A $(= p_2)$.

Application of the above principles will be presented in later sections.

In this brief outline, we have not used the general methods involving the density matrices of the unscattered and scattered beams. Such an approach becomes most useful in those experiments involving more complex spin configurations. The general principles are presented and illustrated in detail by Welton.[12]

TESTS OF FUNDAMENTAL SYMMETRIES

A TEST OF T-INVARIANCE IN THE β-DECAY OF POLARIZED NEUTRONS

A greatly improved experimental upper limit for D, the triple-correlation coefficient in the β-decay of the polarized free neutron, has recently been reported by Steinberg et al.[13] This coefficient appears in the expression for the decay rate in the form[14]

$$D\vec{P}_n \cdot (\vec{p}_e \times \vec{p}_{\bar{\nu}}) \, / \, E_e E_{\bar{\nu}} \tag{9}$$

Here, $\vec{P}_n$ is the neutron polarization and $\vec{p}_e$, $\vec{p}_{\bar{\nu}}$, E_e and $E_{\bar{\nu}}$ are the momenta and energies of the leptons. This expression is odd under time reversal; a non-zero value of D therefore implies a breakdown of T-invariance. The value obtained by Steinberg et al. is

$$D = - \, (1.1 \pm 1.7) \times 10^{-3}$$

which is consistent with T-invariance. The quoted error is largely statistical and is based upon the observation of 5×10^6 events. The phase angle ϕ between the coupling constants g_V and g_A is

$$\phi = 180.14 \pm 0.22^{\circ}$$

In neutron β-decay, the Coulomb interaction is the only important final state interaction and its contribution to D vanishes in a pure V-A theory. Possible weak magnetism effects contribute less than 2×10^{-5}.

The experiment was carried out at the high flux reactor at Grenoble. The cold neutrons had a mean velocity of 1100 m/s and they were polarized by a magnetized curved guide; their mean polarization was $(70 \pm 7)\%$. The beam intensity leaving the polarizer was 10^9 neutron/s and its profile was 5 cm high by 0.6 cm wide. The neutron polarization vector was turned into the beam direction and was periodically changed to be either parallel or anti-parallel to the momentum vector of the neutron.

The magnitude of the neutron momentum may be neglected so that the term (9) can be rewritten

$$D \vec{P}_n \cdot (\vec{p}_p \times \vec{p}_e) / E_e E_{\bar{\nu}} \tag{10}$$

where $\vec{p}_p$ is the momentum of the recoil proton. The experimental geometry was chosen to maximize the triple product as shown:

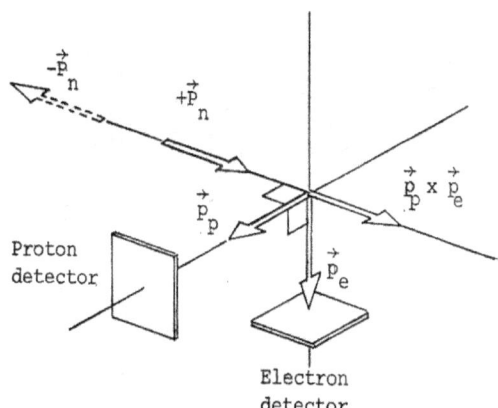

The electrons were detected in a conventional plastic scintillator biased to accept electron energies between 100 and 500 keV. The recoil protons were accelerated to 20 keV and counted in a thin (4000 Å) layer of NaI(Tl). Sixteen time-delayed spectra of coincidences between electron and proton (4 electron detectors, 4 proton detectors and two directions of the incident neutron polarization vector) were recorded. The data were collected during a 2½-month period.

NEUTRON POLARIZATION EFFECTS IN TWO- AND THREE-NUCLEON SYSTEMS

New experiments on $\vec{n}$ - p and $\vec{n}$ - d elastic scattering have been reported recently and all of them are at the forefront of experimental technique. A particularly innovative experiment is that of Brooks and Jones[15,16] whose method opens up interesting possibilities in studies of $\vec{n}$ - p and $\vec{n}$ - d reactions at energies above a few MeV. Before presenting some of their results, a few comments on their method will be useful. In 1964, Tsukada and Kickuchi[17] demonstrated that the scintillation decay of an anthracene crystal excited by 3.7 MeV protons is direction-dependent, relative to the crystal axes. They showed that the fast component is more direction-dependent than the slow component. Brooks and Jones carried out a detailed study of this effect in many different scintillators and, in the course of this work, they invented a new polarimeter suitable for studies of $\vec{n}$ - p and $\vec{n}$ - d interactions. Consider a neutron incident at an angle α with respect to an axis c', normal to an (a,b) plane in the crystal:

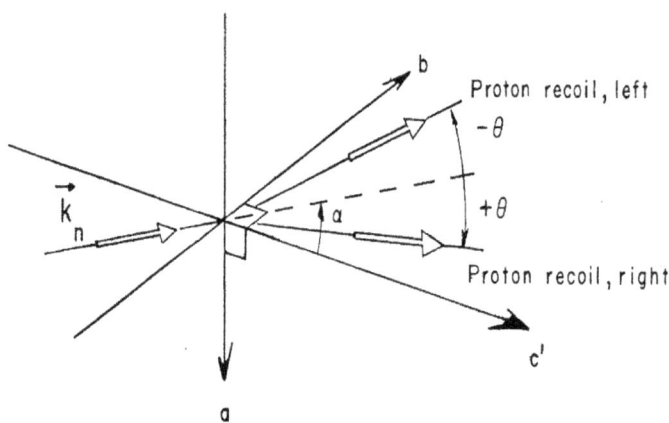

For a proton that recoils through an angle ($-\theta$) towards the b-axis, the values of the integrated light output $\mathcal{L}(-\theta)$ and of the ratio of the slow- to total-light component $\mathcal{S}(-\theta)$ are different from the values obtained when a proton recoils through an angle ($+\theta$) towards the c'-axis. In general,

$$\mathcal{L}(-\theta) < \mathcal{L}(+\theta)$$

and $$\mathcal{S}(-\theta) > \mathcal{S}(+\theta)$$

The measured two-parameter $(\mathcal{L},\mathcal{S})$ data can be analyzed to give the left-right asymmetry in $\vec{n}$-p scattering[18] in an anthracene crystal ($C_{14}H_{10}$) and in $\vec{n}$-d scattering[19] in a deuterated crystal. The $\vec{n}$-d results at neutron energies of 16.4 and 21.6 MeV are shown in Fig. 1.

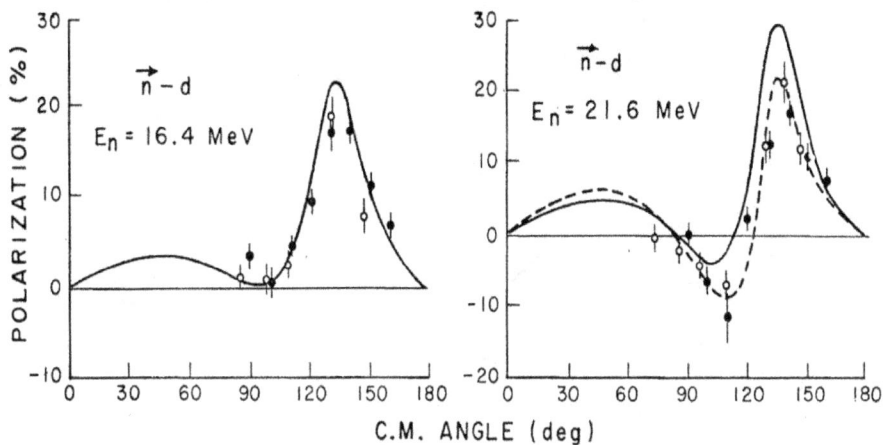

Fig. 1. The $\vec{n}$-d results, at two energies, obtained by Steinbock et al [19] and Morris et al[20] [open circles] compared with the calculations of Pieper[21] [solid curve] and with the p-d measurements[22] [dashed curve].

The work of Morris et al[20] is also shown together with the theoretical calculations of Pieper[21], and with the general trend of the $\vec{p}$-d measurements (the dashed curve).[22] No measurable difference is observed between the polarized neutron and proton induced reactions at these energies. The present status of 3-body theory is discussed in a recent review by Doleschall.[23]

In a demanding experiment, Johnsen et al[24] [see also the contribution to this conference [25]] have measured the spin correlation parameter A_{yy} in $\vec{n}$-$\vec{p}$ scattering at 50 MeV. Their apparatus is shown schematically in Fig. 2.

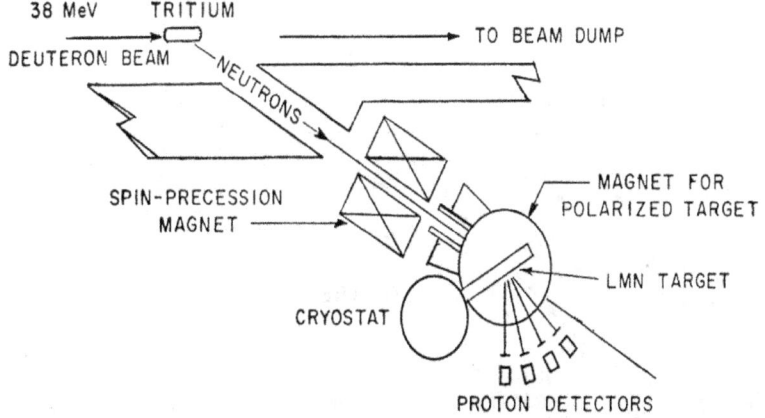

Fig. 2. An arrangement for studying the spin correlation parameter, A_{yy}, in $\vec{n}$-$\vec{p}$ scattering at 50 MeV.[24]

The partially polarized beam from the T-d reaction, is scattered from an aligned LMN proton target. The incident spin direction can be changed with a sole-noidal field. Their results are shown in Fig. 3 where they are compared with the parameters of a recent analysis.[26]

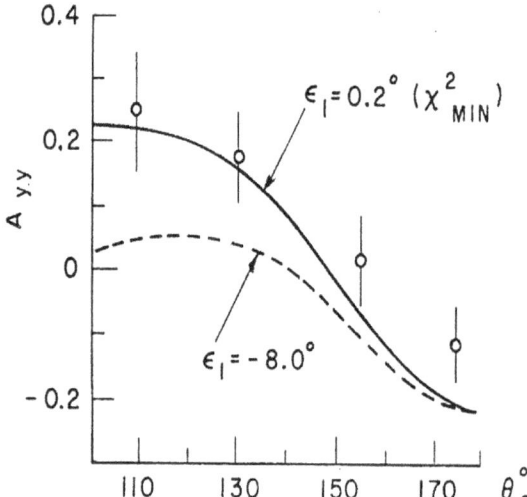

Fig. 3. The results of Johnsen et al[24] compared with the two possible values of ϵ_1 (~ 0 or $-8°$), that gave equally good fits to all n-p data at 50 MeV prior to the present work.[26]

    Neutron triple scattering experiments are notoriously difficult. However, they can provide unique information on the interaction and therefore it is important that they should not be overlooked in future research programs. Ahmed et al[27] have carried out a measurement of the depolarization parameter $D(\theta)$ in $\vec{n}$-d scattering at low energy [see also the contribution to this conference[28]]. Their method is outlined in Fig. 4.

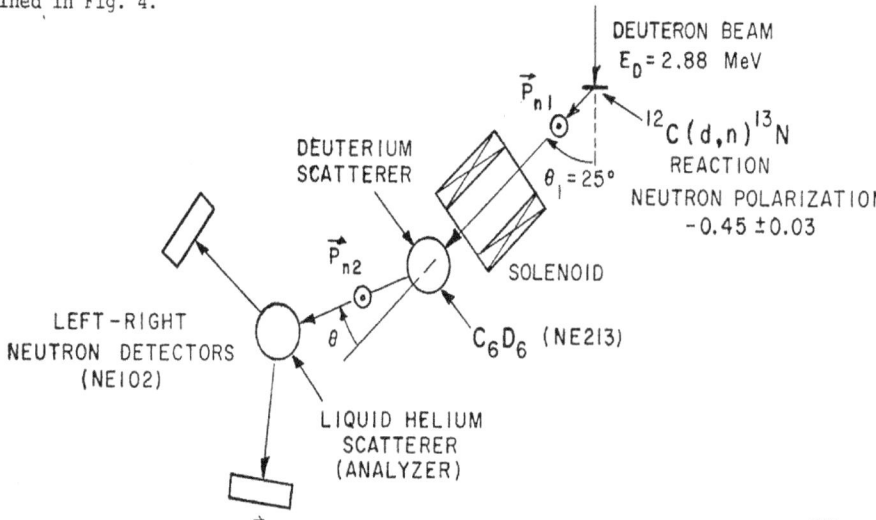

Fig. 4. Apparatus for an $\vec{n}$-d triple-scattering study reported by Ahmed et al.[27]

Their results are compared with theory in Fig. 5.

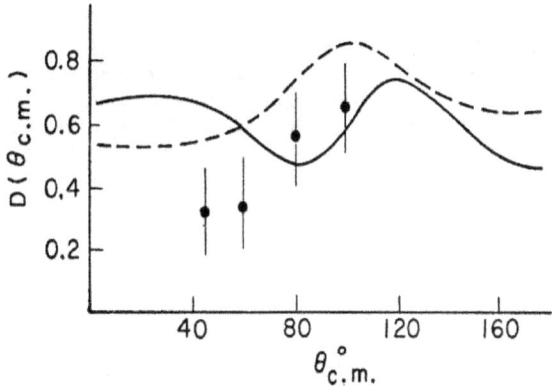

Fig. 5. The results of Ahmed et al[27] compared with two parameterizations of the ERA model.

Clearly, more work remains to be done both experimentally and theoretically before this basic interaction is sufficiently well-understood, even at these low energies.

An important new technique for producing polarized neutrons, particularly in the energy range 10 to 20 MeV, uses the polarization transfer mechanism in D($\vec{d}$,n) and T($\vec{d}$,n) reactions at forward angles.[29] Walter and his colleagues at Duke University have recently reported measurements on ($\vec{n}$,p) scattering at 90° (c.m.) for neutron energies of 13.5 and 16.0 MeV using the D($\vec{d}$,n) reaction as a source.[30] Their results set new standards of precision in neutron polarization studies in the difficult energy region under study; typical statistical accuracy reported is 0.0015. Their results are systematically smaller than the LRLX predictions but are consistent with values calculated with the new phase shifts of Arndt et al.[31]

THE DOUBLE-SCATTERING OF FAST NEUTRONS BY LIGHT NUCLEI

For many years, neutron double-scattering experiments were not considered practicable[32,33] and the first n-^{4}He double-scattering experiment reported[34] did not change the general view. However, in 1972, a program of studies of the polarization of neutrons scattered from light nuclei was successfully initiated at the Yale Electron Accelerator Laboratory, using the double-scattering technique.

The method involves the polarization of an unpolarized flux of neutrons by elastic scattering from ^{12}C. The polarization of the flux scattered at a given angle is measured using true double-scattering in which the polarized flux is scattered again from an identical ^{12}C target at an identical scattering angle. The asymmetry in the doubly scattered flux is measured and, after taking into account the (known) energy-loss at the first scattering, the results are analyzed to give the absolute polarization $\vec{p}$ of the flux.[35,36] Having established the polarization of the source, the analyzing powers of other nuclei can be obtained by replacing the second scatterer with an appropriate target.[37,38,39]

The initial flux of unpolarized neutrons is generated via the (γ,n) reaction in a heavy nucleus and therefore the spectrum is Maxwellian with a maximum intensity at an energy of about 1 MeV. The intensity decreases rapidly at energies above 5 MeV; this is a necessary feature in making measurements of polarization that results from elastic scattering of neutrons in light nuclei.   The neutron energies are measured with good resolution (typically 0.7 ns.m^{-1}). A generalized neutron spin-precession method is used that is well-suited to a continuous energy spectrum of neutrons; this method greatly reduces the systematic errors that would otherwise occur in the experiment.[40,41]

A typical layout of the experiment when used to measure the analyzing power of a light nucleus is shown in Fig. 6. Here, the first reaction angle is 50° and the second scatterer is a cylinder of liquid helium viewed by an array of fast neutron detectors. The observed (source) polarization of neutrons, obtained in a true n-^{12}C double-scattering experiment, is shown in Fig. 7.

The essential points in obtaining the analyzing power when using the generalized spin-precession solenoid are:

The integrated magnetic field required to precess a neutron of measured energy E_π through 180° is

$$\int H.d\ell = 2.37 \times 10^5 \times \sqrt{E_\pi (MeV)} \quad Oe\text{-}cm$$

and the angle of precession, ϕ, of a non-relativistic neutron of measured energy E_ϕ is

$$\phi = \pi\sqrt{E_\pi/E_\phi}$$

The product of the polarization p of the source and the analyzing power A of the second scatterer is

$$pA = \pm(1-R_\pm)/(R_\pm - \cos\phi)$$

where + and – refer to the right and left detector, respectively and

$$R_\pm = \left[N_\pm(H)/N_\pm(0)\right]\left[C(0)/C(H)\right]$$

where $N_\pm(H)$ and $C(H)$ are the corresponding detector count rates and monitor count rate with the field on and $N_\pm(0)$ and $C(0)$ the corresponding rates with the field

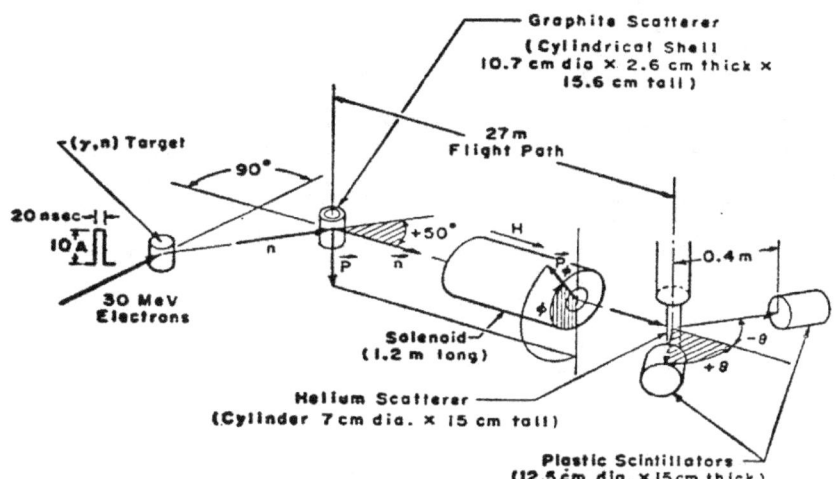

Fig. 6. Schematic diagram of the neutron double-scattering arrangement.

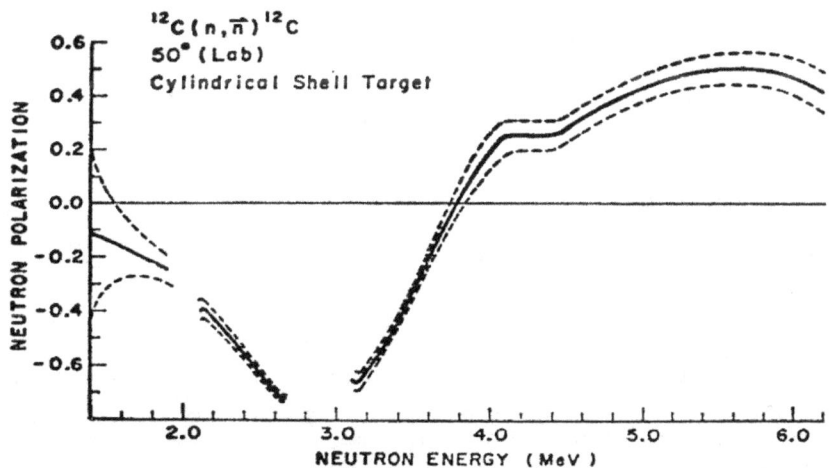

Fig. 7. The observed polarization for neutrons scattered from the graphite target at 50° (lab).

off. It is now straightforward to deduce pA independently of the monitor rates.[41)]

Phase-shift and R-matrix Analyses

The measured polarizations were analyzed using iterative grid search techniques to give definitive sets of phase-shifts and R-matrix parameters. The inclusion of partial waves higher than d-waves did not alter the quality of the fits significantly. Expressions for the differential cross section, polarization and total cross section used are:

$$\sigma(\theta) = (1/k^2) \sum_{L=0}^{4} B_L P_L(\cos\theta)$$

$$\sigma(\theta)p(\theta) = (1/k^2) \sum_{L=1}^{4} C_L \bar{P}_L^1(\cos\theta)$$

$$\sigma_T = (4\pi/k^2) \sum_{L=0}^{2} \left[\ell \sin^2\delta_\ell^- + (\ell+1)\sin^2\delta_\ell^+ \right]$$

where $P_L(\cos\theta)$ and $\bar{P}_L^1(\cos\theta)$ are the Legendre and associated Legendre polynomials. Values for the expression B_L and C_L in terms of phase-shifts have been derived by Blatt and Biedenharn and Simon and Welton.[42,43)]

The elastic scattering of neutrons from spin-zero nuclei is the simplest application of R-matrix theory.[44,45)] Only one channel is open so that

$$R_{\ell J} = \sum_\lambda \gamma_{\lambda \ell J}^2 / (E_{\lambda \ell J} - E)$$

where $\gamma_{\lambda \ell J}^2$ and $E_{\lambda \ell J}$ are the reduced widths and energies, and the states are denoted by λ, and also

$$R_{\ell J} = (f_\ell - B_{\ell J})^{-1}; \quad f_\ell(E) = a u_\ell^{-1}(a)(du_\ell/dr)_a$$

where a is the channel radius, u_ℓ is the radial part of the wave function and $B_{\ell J}$ is the boundary condition. The collision function $U_{\ell J}$ can be expressed in terms of a single, real phase-shift, $\delta_{\ell J}$ thus

$$U_{\ell J} = \exp(2i\delta_{\ell J})$$

The phase-shifts are related to the R-function as follows

$$\delta_{\ell J} = -\phi_\ell + \arctan\left\{ P_\ell R_{\ell J} / \left[1 - R_{\ell J}(S_\ell - B_{\ell J}) \right] \right\}$$

where S_ℓ, P_ℓ and ϕ_ℓ are the well-known shift function, penetrability, and hard-sphere phase-shift. We define the resonance energy E_R as the energy at which the resonant phase-shift is an odd integral multiple of $\pi/2$. The width of the resonance is

$$\Gamma_{\lambda \ell J} = 2P_\ell \gamma_{\lambda \ell J}^2$$

Distant levels are taken into account using the method given in Ref. 46, i.e.

139

$$R_{\ell J}^{\infty} = R_{0\ell J} + R_{\ell J}E$$

A fit was made to the polarization measurements by minimizing the quantity

$$S = \sum_{j=1}^{M} \sum_{i=1}^{N} \frac{\left[P_{cal}(\theta_i,E_j) - P_{exp}(\theta_i,E_j)\right]^2}{\left[\Delta p(\theta_i,E_j)\right]^2}$$

where N is the number of angles (between 4 and 9, depending on the experiment) and M is the total number of energy points used. The optimum R-matrix parameters derived from this procedure, were used to predict the differential and total cross sections, and additional polarizations throughout the entire energy range up to about 5 MeV.

Details of the analysis of the polarization data in the case of $\vec{n}$-^{6}Li scattering are given in a recent paper.[47] This is a complex problem because the target nucleus no longer has spin zero and the (n,α) channel must be properly taken into account. Examples of the measurements and of the analyses of the neutron double-scattering program reported above are shown in Fig. 8. (the observed asymmetry in ^{12}C(50°) - ^{4}He(60°) scattering[39]), in Fig. 9. (the phase-shift analysis of the ^{16}O($\vec{n}$,n)^{16}O reaction, measured at nine angles between 1 and 4 MeV) and in Fig. 10 (the total scattering cross section predicted from an analysis of the polarization data for the ^{16}O($\vec{n}$,n)^{16}O reaction[38]).

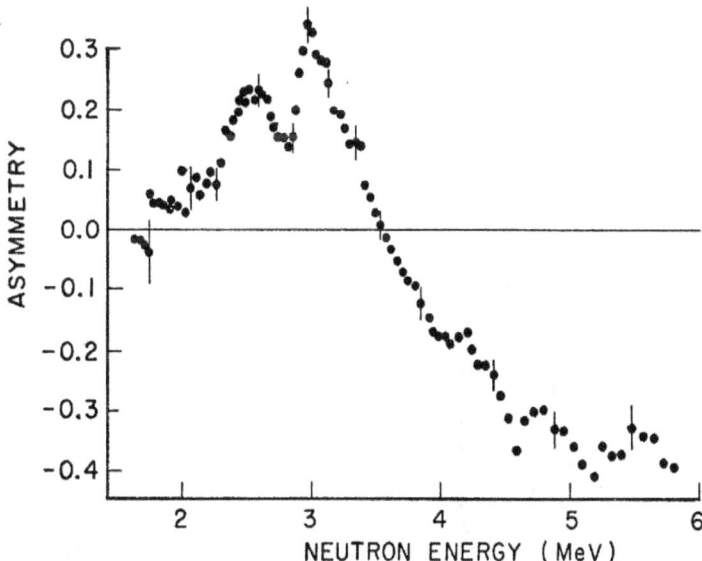

Fig. 8. The observed asymmetry product for ^{12}C(50°) - ^{4}He(60°) neutron double-scattering.[39]

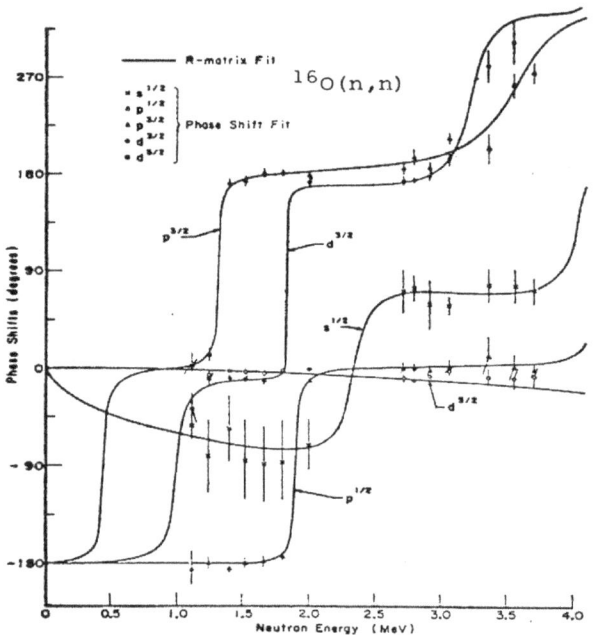

Fig. 9. Phase shifts resulting from the analysis plotted in the first and fourth quadrants. The continuous curve is the R-matrix fit while the discrete points are the result of the phase-shift analysis.

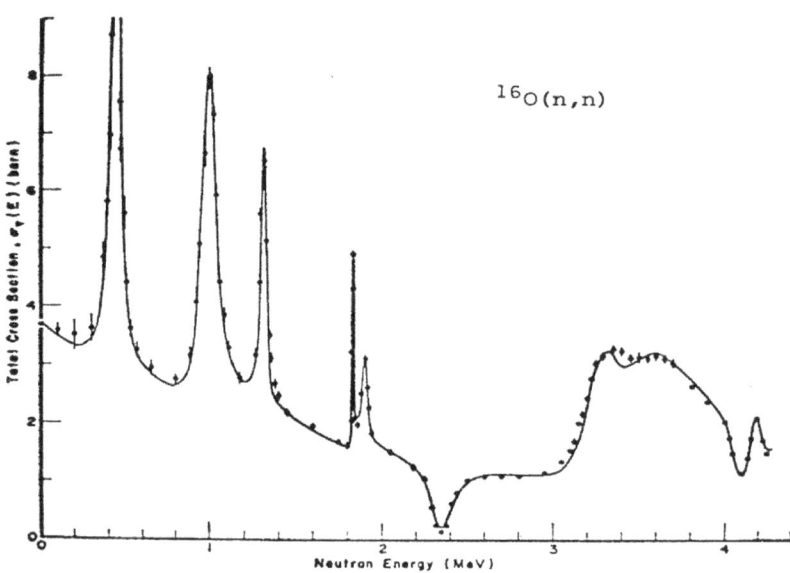

Fig. 10 The R-matrix prediction of the total cross section.

THE TRANSMISSION OF POLARIZED NEUTRONS THROUGH POLARIZED TARGETS

Although it has been known for many years that the measurement of the transmission of polarized neutrons through a polarized target can give the values of the spins of low energy (s-wave) resonances directly, few such measurements have been performed.[48,49,50] This situation is somewhat surprizing in view of the well-established low temperature techniques that form the basis for producing polarized proton targets (the polarizers) and polarized targets. There is, however, one outstanding example of the method, due to Keyworth et al[50] who measured the spins of many resonances in ^{237}Np + n and ^{235}U + n; their work will be discussed in the hope that it may encourage others to take advantage of this powerful technique.

If a beam of polarized neutrons (polarization p_n) is passed through a polarized target (polarization p_T) and the directions of the vectors $\vec{p}_n$ and $\vec{p}_T$ are parallel, then the transmission $T_J^{\uparrow\uparrow}$ is given by[51]

$$T_J^{\uparrow\uparrow} = e^{-n\sigma_J} \{\cosh(\rho_J p_T n\sigma_J) - p_n \sinh(\rho_J p_T n\sigma_J)\}$$

where n is the number of nuclei/cm^2 in the target, σ_J is the unpolarized cross section and

$$J = I \pm 1/2,$$

is the spin of the resonance, and I is the target spin.

If we write $J_+ = I + 1/2$ and $J_- = I-1/2$, the values of ρ_J are

$$\rho_{J_+} = I/(I+1) \text{ and } \rho_{J_-} = -1$$

The transmission $T_J^{\downarrow\uparrow}$, corresponding to the vectors $\vec{p}_n$ and $\vec{p}_T$ being anti-parallel, is

$$T_J^{\downarrow\uparrow} = e^{-n\sigma_J} \{\cosh(\rho_J p_T n\sigma_J) + p_n \sinh(\rho_J p_T n\sigma_J)\}$$

For a $J_+ = I + 1/2$ state, the difference in transmission for parallel and anti-parallel polarization vectors is therefore

$$T_{J_+}^{\uparrow\uparrow} - T_{J_+}^{\downarrow\uparrow} = \Delta T_{J_+} = -2\,p_n e^{-n\sigma_{J_+}} \cdot \sinh\left[\left(\frac{I}{I+1}\right) p_T n\sigma_{J_+}\right] \qquad (11)$$

which is always negative.
If, however, a state has $J_- = I - 1/2$, the difference in transmission is

$$T_{J_-}^{\uparrow\uparrow} - T_{J_-}^{\downarrow\uparrow} = \Delta T_{J_-} = +2p_n e^{-n\sigma_{J_-}} \cdot \sinh\left[p_T n\sigma_{J_-}\right] \qquad (12)$$

which is always positive.

Keyworth <u>et al</u> [50]   used a high intensity, unpolarized, pulsed neutron
beam from ORELA and obtained a polarized neutron beam by passage through a
polarized LMN target (p_n = 0.55). The direction of the vector $\vec{p}_n$ could be re-
versed with a   magnetic field. The transmission of this beam through
a polarized ^{237}Np (or ^{235}U) target (p_T = 0.2) was measured for the two incident
polarization directions. In addition, the yield of fast fission neutrons was
measured in an array of detectors around the (second) polarized target. The clear-
cut determination of resonance spins in ^{237}Np + n, using Eqs. (11 and 12) is shown
in Fig. 11 (In this case, I = 5/2, therefore J_+ = 3 and J_- = 2)

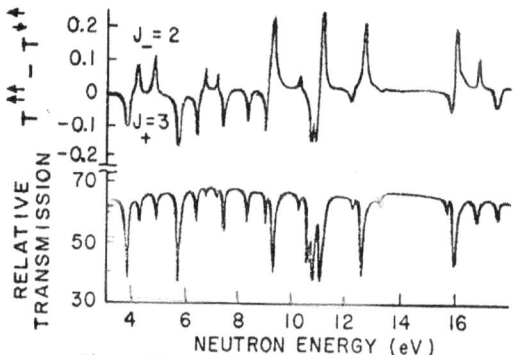

Fig. 11. Values of $T^{\uparrow\uparrow} - T^{\downarrow\uparrow}$ and the relative transmission for ^{237}Np + n showing
the clear-cut determination of the resonance spins.[50]

A most interesting conclusion from their experiment is that the fine-structure
resonances in a given intermediate-structure group in the fission yield have the
same spin.

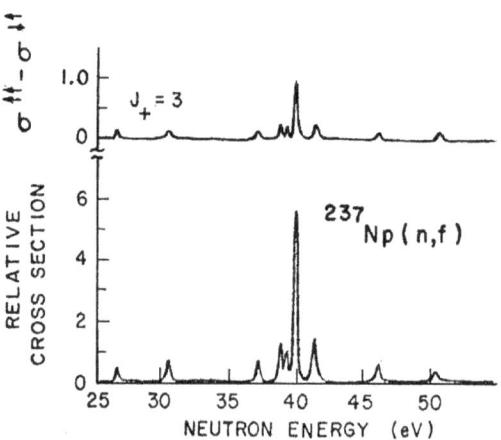

Fig. 12. Measurement of the spins of resonances in the fission of ^{237}Np in the
vicinity of the intermediate structure at 40 eV.[50]

NEUTRON POLARIZATION EFFECTS IN NUCLEAR PHOTO-DISINTEGRATION

The shell model forms the basis of nuclear structure theory so that any attempt to provide a more quantitative understanding of the model is a matter of fundamental importance. The interaction between photons and nuclei affords a sensitive probe of certain features of nuclear structure. At photon energies below about 30 MeV, the interaction is predominantly electric dipole in character, so that the incident photon only excites a limited number of states which have the correct spins and parities (consistent with the addition of one unit of angular momentum and a change in parity). The following qualitative description of the photon-nucleus interaction illustrates the major points of the problem. Consider a light nucleus with a ground state configuration which is well-described by the shell model. The most straightforward examples are those nuclei which have closed shells of neutrons and protons eg: ^{16}O, ^{40}Ca and ^{208}Pb.

An El photon excites a nucleon into a higher energy state; if the photon energy E_γ is sufficiently high then the nucleon becomes unbound and is emitted (in light nuclei, $E_\gamma \gtrsim 15$ MeV for neutron emission to occur). A 1 particle -1 hole (1p-1h) state is thereby created; in the case of El absorption in ^{16}O, five such 1p-1h states are possible. The final "electric dipole states" Ψ_D^* are considered to be linear combinations of the five base states:

$$\Psi_D^* \propto \sum_{i=1}^{5} c_i \phi_i$$

If the combination happens to be coherent then a strong transition will be observed. In ^{16}O, there are two such transitions which, between them account for more than 90% of the total dipole absorption strength. These states, at 22 and 25 MeV were predicted, in a calculation of this type, by Elliott and Flowers[52] in 1957 and were subsequently observed in the reactions $^{15}N(p,\gamma_0)^{16}O$, $^{16}O(e,e',p_0)^{15}N$, and $^{16}O(\gamma,n_0)^{15}O$ between 1959 and 1962.

A signigicant test of the shell model used in the Elliott-Flowers calculation requires a determination of the amplitudes c_i which are associated with the five base states. It will be shown that it is not possible to answer these questions simply by measuring the angular distributions of the outgoing photonucleons - a measurement of their differential polarizations is also necessary.

In general, the angular distribution of the photonucleons is of the form:

$$\frac{d\sigma}{d\Omega} = \sum_{L=0}^{N} A_L P_L (\cos\theta)$$

where $P_L(\cos\theta)$ is the Legendre polynomial of order L and N is the maximum allowed value of L (consistent with the conservation of angular momentum).

In the present case of El absorption in ^{16}O, the differential cross section for photoneutrons emitted to the ground state of ^{15}O has the form:

$$\frac{d\sigma}{d\Omega} = \frac{3}{16} \lambdabar_\gamma^2 \{2(a_s^2 + a_d^2) + (2\sqrt{2}\, a_s a_d \cos\Delta_{sd} - a_d^2) P_2(\cos\theta)\} \qquad (13)$$

Here, a_s and a_d are the real magnitudes of the s- and d- wave emission amplitudes respectively and $\Delta_{sd} = \delta_s - \delta_d$, where δ_s and δ_d are their respective phases. The only reaction matrix elements which contribute in this case are:

$$\langle \ell = 0, s = 1, \alpha' |R^{1^-}|El\alpha\rangle \equiv a_s e^{i\delta_s} \qquad (14)$$

and
$$\langle \ell = 2,\ s = 1,\ \alpha' | R^{1^-} | E1\alpha \rangle \equiv a_d e^{i\delta_d} \tag{15}$$

where s is the channel spin and α, α' specify the ground state ($J^\pi = 0^+$) and excited states ($J^\pi = 1^-$) respectively. The coefficients, A_L are given by:

$$A_o \sim a_s^2 + a_d^2$$

and
$$A_2 \sim (a_s a_d \cos\Delta_{sd} - a_d^2)$$

Now $4\pi A_o$ is the total cross section and is insensitive to interference effects between the different components (the $\ell = 0$ and $\ell = 2$ partial waves associated with the outgoing nucleons). Although the A_2 - coefficient contains an interference term, it is nonetheless finite if the state only emits d-wave ($\ell = 2$) nucleons. The differential polarization of photonucleons has the form

$$\frac{d\vec{P}}{d\Omega} = \hat{k} \sum_{L=1}^{N} B_L \bar{P}_L (\cos\theta) \tag{16}$$

where $\hat{k}$ is a unit vector normal to the scattering plane and $\bar{P}_L(\cos\theta)$ is the associated Legendre polynomial. The significant difference between the expressions for the angular distribution and polarization of photonucleons is the absence of the L = 0 term in the summation in Eq. (16). This reflects the fact that any polarization produced is due to interference effects between different channels. In the present example, it is found that:

$$\frac{d\vec{P}}{d\Omega} = k\lambda_\gamma^2 (0.205\ a_s a_d \sin\Delta_{sd})\ \bar{P}_2^1 (\cos\theta) \tag{17}$$

which shows that the differential polarization is zero (for all values of θ) if either $a_s = 0$ or $a_d = 0$.

Another important point emerges from the expression for $d\vec{P}/d\Omega$ given in Eq. (17): the associated Legendre function of second order is:

$$\bar{P}_2^1 (\cos\theta) = -\sqrt{\frac{15}{16}} \sin 2\theta \tag{18}$$

which means that, at a reaction angle of $\theta = 90^o$, the polarization $d\vec{P}/d\Omega = 0$. Conversely, the appearance of any polarization at $\theta = 90^o$ is clear evidence of the intrusion of M1 or E2 multipoles in the absorption process.

With these points in mind, three groups[53,54,55] have measured the polarization of photoneutrons from a number of nuclei at appropriate angles. The most recent work involves studies of the reactions $d(\gamma,\vec{n})p$ [56], $^{16}O(\gamma,\vec{n})^{15}O$ [57] and $^{208}Pb(\gamma,\vec{n})^{207}Pb$ [58], using the method developed at Yale in which an intense pulsed source of electrons produces a bremsstrahlung photon spectrum (in a tungsten converter) with a maximum energy set to avoid exciting non-ground state transitions in ^{16}O and ^{208}Pb. The neutron energies are determined with good resolution (<1 ns.m^{-1}) and the neutron polarization determined by measuring the left-right asymmetry in scattering from a suitable light nucleus (^{24}Mg at neutron energies between 0.1 and 0.5 MeV, ^{16}O between 0.3 and 1.5 MeV, ^{12}C between 1 and 10 MeV and ^4He between 1 and 20 MeV).

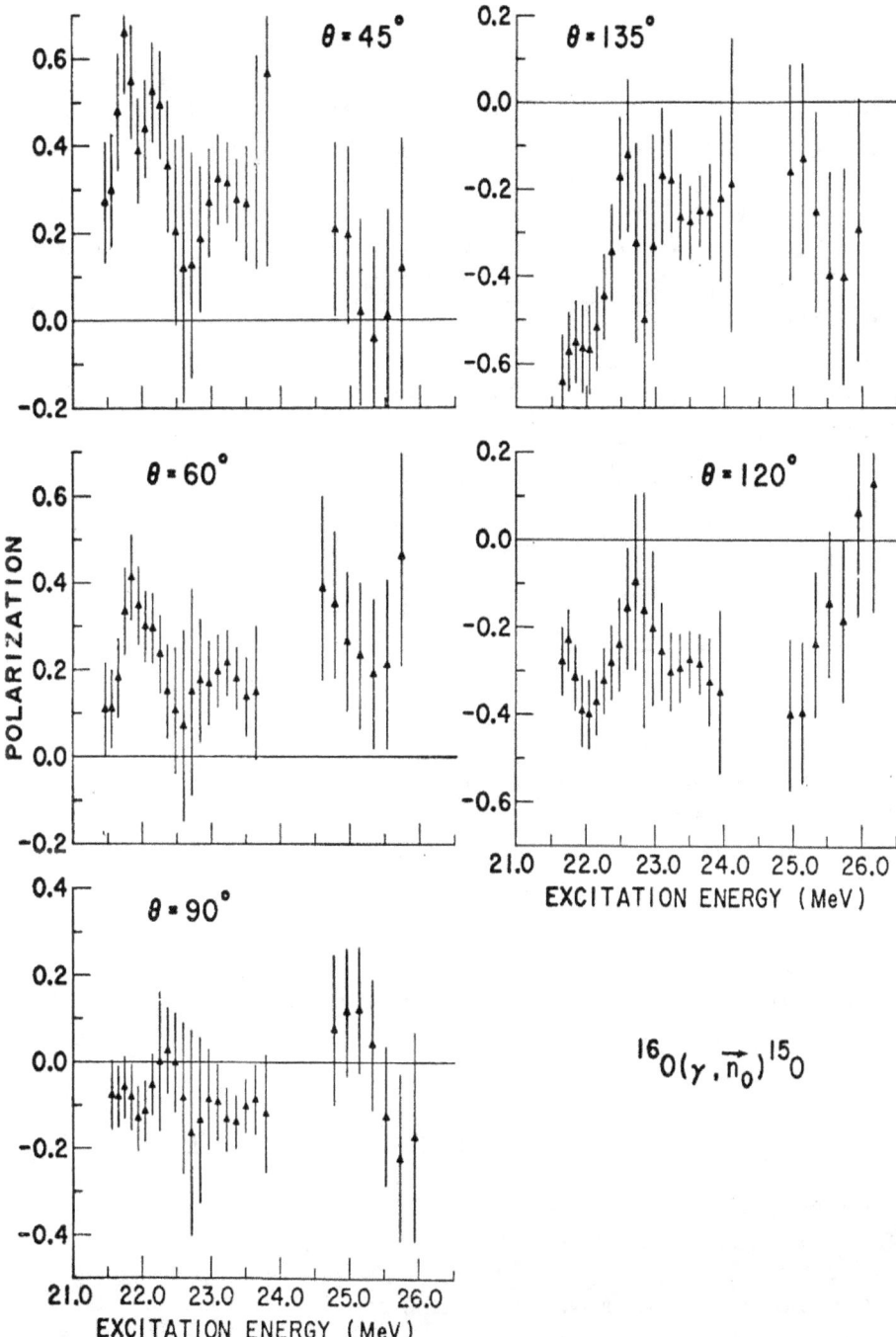

Fig. 13. The angular distribution of polarization of photoneutrons in the
$^{16}O(\gamma, \vec{n}_0)^{15}O$ reaction in the region of the main dipole states.[57]

146

REFERENCES

[1] R. K. Adair, S. E. Darden and R. E. Fields, Phys. Rev. 96 (1954) 503.

[2] A. M. Baldin, V. I. Goldanskii and I. L. Rozental', Kinematics of Nuclear Reactions, Oxford Univ. Press, 1961.

[3] M. I. Shirokov, Zhur. Eksp. i Teoret. Fiz. 32 (1957) 1022.

[4] E. M. Henley, Ann. Rev. Nucl. Science 19 (1969) 367.

[5] H. A. Weidenmüller, Polarization Phenomena in Nuclear Reactions, Univ. of Wisconsin Press, 1971.

[6] L. Wolfenstein, Ann. Rev. Nucl. Science 6 (1956) 43.

[7] W. Haeberli, Fast Neutron Physics, Vol. II, Interscience, 1963.

[8] H. H. Barschall, 2nd Inter. Polarization Conf. Karlsruhe, 1965.

[9] R. L. Walter, 4th Inter. Symposium on Polarization Phenomena in Nuclear Reactions, 1976 (in press).

[10] N. F. Mott and H. S. W. Massey, The Theory of Atomic Collisions, Oxford Univ. Press, 1965.

[11] The "Basel" Convention.

[12] T. A. Welton, Fast Neutron Physics, Vol. II, Interscience, 1963.

[13] R. I. Steinberg, P. Liaud, B. Vignon and V. W. Hughes, Phys. Rev. Lett. 33 (1974) 41.

[14] J. D. Jackson, S. B. Treiman and H. W. Wyld Jr., Phys. Rev. 106 (1957) 517.

[15] F. D. Brooks and D. T. L. Jones, Nucl. Inst. 121 (1974) 69.

[16] D. T. L. Jones and F. D. Brooks, Nucl. Phys. A222 (1974) 79.

[17] D. Tsukada and S. Kickuchi, Nucl. Inst. 17 (1964) 286.

[18] F. D. Brooks and D. T. L. Jones, Nucl. Inst. 121 (1974) 77.

[19] M. Steinbock, F. D. Brooks and I. J. van Heerden, 4th Inter. Symposium on Polarization Phenomena in Nuclear Reactions, 1976 (in press).

[20] C. L. Morris, R. Rotter, W. Dean, and S. T. Thornton, Phys. Rev. C9 (1974) 1687.

[21] S. C. Pieper, Nucl. Phys. A193 (1972) 519.

[22] J. C. Faivre, D. Garreta, J. Jungerman, A. Papineau, J. Sura and A. Tarrata, Nucl. Phys. A127 (1969) 169.

[23] P. Doleschall, 4th Inter. Symposium on Polarization Phenomena in Nuclear Reactions, 1976 (in press).

[24] S. W. Johnsen, F. P. Brady, N. S. P. King and M. W. McNaughton, ibid.

[25] J. L. Romero, F. P. Brady, N. S. P. King and M. W. McNaughton, contribution to this conference.

[26] J. Binstock and R. Bryan, Phys. Rev. D9 (1974) 2528.

[27] M. Ahmed, D. Bovet, P. Chatelain and J. Weber, 4th Inter. Symposium on Polarization Phenomena in Nuclear Reactions, 1976 (in press).

28) D. Bovet, P. Chatelain, S. Jaccard, Y. Onel, J. Piffaretti, R. Vinet and J. Weber, contribution to this conference.

29) J. E. Simmons, W. B. Broste, G. P. Lawrence, J. L. McKibben and G. C. Ohlsen, Phys. Rev. Lett. 27 (1971) 113.

30) W. Tornow, P. W. Lisowski, R. C. Byrd, S. E. Skubic and R. L. Walter, B.A.P.S. 21 (1976) 636.

31) R. A. Arndt, private communication to R. L. Walter and M. H. MacGregor, R. A. Arndt and R. M. Wright, Phys. Rev. 182 (1969) 1714.

32) W. Haeberli, Fast Neutron Physics, Vol. II, Interscience, 1963.

33) J. M. Daniels, Oriented Nuclei: Polarized Targets and Beams, Academic Press, 1965.

34) R. B. Perkins and C. Glashausser, Nucl. Phys. 60 (1964) 433.

35) R. J. Holt, F. W. K. Firk, R. Nath and H. L. Schultz, Phys. Rev. Lett. 28 (1972) 114.

36) R. J. Holt, F. W. K. Firk, R. Nath and H. L. Schultz, Nucl. Phys. A213 (1973) 147.

37) G. T. Hickey, F. W. K. Firk, R. J. Holt, R. Nath and H. L. Schultz, Phys. Lett. 47B (1973 348.

38) G. T. Hickey, F. W. K. Firk, R. J. Holt and R. Nath, Nucl. Phys. A225 (1974) 470.

39) J. E. Bond and F. W. K. Firk, Nucl. Phys. A258 (1976) 189.

40) P. Hillman, G. H. Stafford and C. Whitehead, Nuovo Cimento, 4 (1956) 67.

41) R. Nath, F. W. K. Firk, R. J. Holt and H. L. Schultz, Nucl. Inst. 98 (1972) 385.

42) J. M. Blatt and L. Biedenharn, Rev. Mod. Phys. 24 (1952) 258.

43) A. Simon and T. A. Welton, Phys. Rev. 90 (1953) 1036.

44) E. P. Wigner and L. Eisenbud, Phys. Rev. 72 (1947) 29.

45) A. M. Lane and R. G. Thomas, Rev. Mod. Phys. 30 (1958) 257.

46) F. W. K. Firk, J. E. Lynn and M. C. Moxon, Proc. Phys. Soc. (Lond) 82 (1963) 477.

47) R. J. Holt, F. W. K. Firk, G. T. Hickey and R. Nath, Nucl. Phys. A237 (1975) 111.

48) H. Postma, H. Marshak, V. L. Sailor, F. J. Shore and C. A. Reynolds, Phys. Rev. 125 (1962) 979.

49) F. L. Shapiro, Nuclear Structure Study with Neutrons, North-Holland, 1966.

50) G. A. Keyworth, J. R. Lemley, C. E. Olsen, F. T. Seibel, J. W. T. Dabbs and N. W. Hill, Phys. Rev. C8 (1973) 2352.

51) A. Stolovy, Phys. Rev. 118 (1960) 211, and 134B (1964) 68.

52) J. P. Elliott and B. H. Flowers, Proc. Roy. Soc. A242 (1957) 57.

53) W. Bertozzi, P. T. Demos, S. Kowalski, C. P. Sargent, W. Turchinetz, R. Fullwood and J. Russell, Phys. Rev. Lett. 10 (1963) 106.

148

54) G. W. Cole Jr., F. W. K. Firk and T. W. Phillips, Phys. Lett. 30B (1969) 91 and R. Nath, F. W. K. Firk and H. L. Schultz, Nucl. Phys. A194 (1972) 49.

55) R. J. Holt and H. E. Jackson, Phys. Rev. Lett. 36 (1976) 244.

56) L. Drooks, F. W. K. Firk, H. L. Schultz and R. J. Holt, B.A.P.S. 91 (1976) 534.

57) R. Nath, Y-H. Chiu and F. W. K. Firk, to be published.

58) R. J. Holt, R. M. Laszewski and H. E. Jackson, to be published.

59) H. F. Glavish, Inter. Conf. on Photonuclear Reactions and Applications, Vol. II, Lawrence Livermore Laboratory, Univ. of California, 1973.

60) D. B. C. B. Syme and G. I. Crawford, ibid.

61) R. J. Holt, private communication, 1976 and refs. 55 and 58.

62) L. Drooks, Ph.D. Thesis, Yale University (1976).

63) J. Blomquist, Phys. Lett. 32B (1970) 1.

64) D. O. Riska and G. E. Brown, Phys. Lett. 38B (1972) 193.

65) E. Hadjimichael, Phys. Lett. 46B (1973) 147 and private communication.

66) H. Arenhövel, W. Fabian and H. G. Miller, Phys. Lett. 52B (1974) 303.

67) F. Partovi, Ann. of Phys. 27 (1964) 79.

68) R. D. Nunemaker, Ph.D. Thesis, Yale University, 1968.

69) J. M. Potter, J. D. Bowman, C. F. Hwang, J. L. McKibben, R. E. Mischke, D. E. Nagle, P. G. Debrunner, H. Frauenfelder and L. B. Sorenson, Phys. Rev. Lett. 33 (1974) 1307.

NEUTRON TIME-OF-FLIGHT SPECTROMETERS

F. W. K. FIRK

Department of Physics, Yale University, New Haven, Conn., U.S.A.

1. Introduction

There are many reasons why it is difficult to measure neutron energies with high resolution and reasonable efficiency. Among them are the lack of net electric charge on the neutron, which rules out all forms of electro-magnetic and direct ionization spectrometers, the difficulties of producing well-collimated, intense fluxes of neutrons within narrow ranges of energy, the high levels of background radiation that frequently accompany neutron-producing reactions, and the need to use indirect methods for detecting neutrons. In spite of these drawbacks, several methods of neutron spectroscopy have been developed including (1) the measurement of proton recoil energies, (2) the measurement of the energy release in a known nuclear reaction, (3) diffraction in crystals at very low energies, and (4) time-of-flight measurements. The latter has emerged as the leading method; it is widely used throughout nuclear and particle physics, and in areas of solid state physics.

Only three years after Chadwick's discovery of the neutron in 1932, Dunning et al.[1]) developed a neutron velocity selector to measure the low-energy part of the spectrum of neutrons from a radon–beryllium source, surrounded by a moderating material (paraffin wax). A two–shutter system of rotating metal discs, each with sectors of cadmium, was used to generate pulses of neutrons from the continuous flux from the source, and to select the neutrons into groups of definite speed. The energy of the neutrons selected depended on the distance between the discs, on their speed of rotation, and on their relative phase. For the first time, a peak in the neutron spectrum, characteristic of the Maxwellian shape expected from a thick neutron moderator was observed. This work formed the basis for the general development of neutron velocity selectors, or time-of-flight spectrometers, using high-speed mechanical rotors to generate pulses of neutrons of short (microsecond) duration from nuclear reactors[2]).

In 1938, Alvarez[3]) reported a study of the neutron absorption of boron at very low energy that used the time-of-flight technique to discriminate between fast (MeV) neutrons that formed an unwanted background, and the low energy (eV) neutrons of interest. The deuteron beam of the 36″ cyclotron at Berkeley struck a beryllium target and the resulting flux of low energy neutrons was moderated with a suitable material. The deuteron beam was effectively pulsed by modulating the plate voltages of the power amplifiers for the dees of the cyclotron. According to Alvarez[3]) the possibility of using neutron time-of-flight spectroscopy as a general method had been discussed by the members of the cyclotron group at Berkeley as early as 1936. During the two decades following the work of Alvarez, pulsed neutron sources became available using reactions induced by protons, deuterons and alpha particles from Cockcroft–Walton generators[4]), cyclotrons[5]), and Van de Graaff accelerators[6,7]), and by photons obtained from linear electron accelerators[8,9]) and from betatrons[10]).

Before discussing the time-of-flight method in detail, it will help to place it in its proper perspective by outlining some of the characteristics of the first three methods listed above.

1.1. RECOIL COUNTER SPECTROMETERS

At neutron energies above about 100 keV, it is possible to detect neutrons with reasonable efficiency by detecting recoiling nuclei (generally protons). The main types of counter involve gas recoil detection, recoil detection in scintillators or in photographic plates.

Both the gas recoil and scintillation recoil counter are not suitable for spectrometers which require high resolving power and high efficiency. This is due to the fact that the energy spectrum of recoil protons which results from an incident monoenergetic neutron source is not monoenergetic, but has a continuous energy distribution. The pulse height from such a system is no longer uniquely related to the neutron energy; it is therefore necessary to derive a differential energy spectrum from an integral pulse height distribution. An improvement can be made by observing only those protons that recoil in a particular direction (usually close to 0°) but this imposes severe limitations on the efficiency of the spectrometer.

The use of photographic plates to observe neutron induced reactions has several advantages compared with the other techniques; these include a permanent track record left by the recoil nucleus, the ability to make accurate measurements on the range, angle and ionization of the recoiling nuclei, and the relative insensitivity to photon backgrounds. Their main disadvantages are associated with difficulty in detecting neutrons with energies below about 0.5 MeV (due to the fact that the minimum track length necessary for identification is about 3 μm) and with a lack of resolution. It is difficult to measure neutron energies with uncertainties of less than 100 keV and, although this may be satisfacotry for energies above 20 MeV, it is rarely adequate at lower energies.

1.2. MEASUREMENT OF THE ENERGY RELEASE IN A KNOWN NUCLEAR REACTION

The low efficiency and difficulty in achieving high resolution with recoil spectrometers has resulted in a number of spectrometers based upon a neutron induced reaction of the type $n + A \rightarrow B + C + Q$. In this reaction, the product particles are accompanied by an energy release Q. This energy, plus the kinetic energy of the incident neutron appears as kinetic energy of the product particles, provided they are not in excited states. Although the individual energies of B and C vary with their angles of emission, their combined energy is always the sum of Q and the incident neutron energy. The neutron energy is therefore obtained from a measurement of the total energy released.

The reaction must satisfy several requirements if it is to be useful in a neutron spectrometer. For example, the cross section must be (sufficiently) large, and a smooth function of the neutron energy. There must be few alternative reactions possible (these alternatives include those in which one or both of the reactions products is left in an excited state). If the low energy range is to be covered then Q must be positive. The material containing the target nuclei (A) must be in a form which is suitable for use in, or with, an energy sensitive detector.

There are no neutron induced reactions which completely satisfy all of these requirements. The choice is limited to the light nuclei ^{3}He, ^{6}Li, ^{10}B

and ^{14}N. Of these, the most favorable is the ^{3}He-reaction:

$$n + {}^3\text{He} \rightarrow p + {}^3\text{H} + 770 \text{ keV}.$$

The efficiency of the ^{3}He-counter depends primarily upon the amount of the gas that can be incorporated into the sensitive volume.

The intrinsic spread in pulse height from a proportional counter is a function of the mean number of ions initially liberated (i.e. excluding those produced in the avalanche). In ^{3}He, the intrinsic resolution is about $1\frac{1}{2}\%$ at 1 MeV. Other factors, not fundamental in origin also have an appreciable effect on the resolution. These include well-known effects such as mechanical imperfections, variations in the high voltage applied to the counter and impurities in the gas filling. Taking all these factors into account, it is possible to achieve an overall resolution between 3 and 5% at neutron energies above 1 MeV.

1.3. DIFFRACTION IN CRYSTALS AT LOW ENERGIES

High resolution neutron spectrometers have been in use for many years that make use of the reflection of an intense well-collimated beam of neutrons (usually from a reactor) at the Bragg angles associated with the atomic planes of a crystal. The standard relationship holds:

$$n\lambda = 2d \sin \theta,$$

where n is an integer, λ is the De Broglie wavelength of the neutron, d is the spacing between crystal planes and θ is the angle between the direction of the incident neutron beam and the reflecting plane.

Neutrons of wavelength λ, reflected at an angle θ, are collimated and detected in a ^{10}BF$_3$ counter or a ^{6}Li-loaded glass scintillator. The counter can be rotated to allow a variation in the angle θ, and hence the spectrometer can cover a range of neutron energies.

The resolving power of such a spectrometer is largely governed by the uncertainties in defining the angle θ. In addition, there is an intrinsic uncertainty introduced by imperfections in the crystal structure itself. For a particular experiment, the uncertainty $\delta\theta$ in θ is related to the uncertainty $\delta\lambda$ in λ by $n\delta\lambda \simeq 2d\delta\theta$ (for θ close to 0°), and the uncertainty δE in the neutron energy E varies as $E^{\frac{3}{2}}$; we shall see that this is the same dependence on neutron energy as that obtained in a time-of-flight spectrometer in which δE is due entirely to a timing uncertainty δt.)

Crystal spectrometers have been used with great success up to energies of about 10 eV. Their performance deteriorates above this energy due to a loss in counting rate as the neutron energy increases. The neutron flux from a reactor decreases as $1/E$ and, furthermore, the crystal reflectivity varies as $1/E$. The decrease in the efficiency of the neutron detector with increasing energy must also be considered (this problem is, of course, common to most forms of low-energy neutron spectrometers).

Resolutions of less that 0.2% at 1 eV have been achieved using reflection from the $22\bar{4}2$ planes of a beryllium crystal. Clearly, a small value of the spacing d is desirable; beryllium is found to be suitable because of the high density of unit cells and, furthermore, only a small amount of thermal lattice disorder is observed.

The main advantages of a crystal spectrometer stem from the relatively high neutron intensities available in small cross sectional areas, so that exotic materials may be readily studied, and from the time-independence of the neutron source which means that the time-characteristics of the neutron detector are unimportant.

2. Principles of the method

2.1. RELATIVISTIC NEUTRONS

The speed v_n of a neutron can be determined by measuring the time t_n that it takes to travel a measured distance l in free space. The kinetic energy E_n of the neutron can then be deduced knowing its speed, rest energy E_0 ($= 939.553$ MeV) and the speed of light c, using the familiar result of special relativity:

$$E_n = E_0[(1-v_n^2/c^2)^{-\frac{1}{2}} - 1]$$
$$= E_0[(1-l^2/t_n^2 c^2)^{-\frac{1}{2}} - 1]. \quad (1)$$

If the units of energy are MeV, and those of length and time are m and ns, then*

$$E_n = 939.553 \left[(1 - 11.126496\, l^2/t_n^2)^{-\frac{1}{2}} - 1\right] \text{MeV}. \quad (2)$$

It is frequently useful to rearrange this equation to obtain the ratio t_n/l for a given energy E_n:

$$\frac{t_n}{l} = \frac{3.3356404}{\left[1 - \left(\dfrac{939.553}{E_n+939.553}\right)^2\right]^{\frac{1}{2}}} \frac{\text{ns}}{\text{m}}. \quad (3)$$

* The values of the constants are taken from the review article of Taylor et al.[11]).

TABLE 1

Typical values of t_n/l for several energies.

E_n (MeV)	t_n/l (ns/m)
1	72.355
5	32.461
10	23.044
100	7.795

Typical values of this quantity are given in table 1.

These values play an important part in helping to establish some of the basic requirements of a spectrometer.

2.2. NON-RELATIVISTIC NEUTRONS

At energies below about 1 MeV, the non-relativistic approximation to eq. (3) is often adequate:

$$\left(\frac{t_n}{l}\right)_{\text{N.R.}} = \left(\frac{E_0}{2E_n c^2}\right)^{\frac{1}{2}} = \frac{72.298}{\sqrt{E_n}} \frac{\text{ns}}{\text{m}}.$$

At $E_n = 1$ MeV, the difference between the relativistic and non-relativistic calculations of (t_n/l) is 0.057 ns/m.

In the eV-region, it is the custom to use units of μs/m for the quantity (t_n/l); a 1 eV neutron takes 72.3 μs to travel 1 m.

2.3. THEORETICAL FACTORS

Before discussing the various experimental factors effecting the resolving power of a particular neutron time-of-flight spectrometer, it will be useful to outline the known theoretical factors involved in the study of neutron resonance reactions so that the necessary performance of such a spectrometer can be specified.

In nuclei with mass number $A \gtrsim 20$, neutron interactions below an energy of about 1 MeV are characterized by the appearance of sharp, closely spaced resonances associated with quasi-stationary states of the compound system[12]). Examples of neutron resonances observed in the total cross sections of natural sodium[13]) and iodine[14]) are shown in figs. 1 and 2.

The neutron total cross section $\sigma_{n,T}(E)$ is obtained by measuring the transmission $T(E)$ of monoenergetic neutrons of energy E through a uniform slab of material containing n nuclei per cm^2, thus:

542   F. W. K. FIRK

$$\sigma_{n,T}(E) = \frac{1}{n} \ln\left[\frac{1}{T(E)}\right].$$

The form of the total cross section in the region of an isolated s-wave resonance is given by the Breit–Wigner[15] expression

$$\sigma(x) = \frac{\sigma_0}{1+x^2} + \tan 2ka \cdot \frac{\sigma_0 x}{1+x^2} + \sigma_{pot},$$

where

$$x = 2(E-E_R)/\Gamma,$$

$$\sigma_0 = \frac{4\pi}{k^2} g \frac{\Gamma_n}{\Gamma} \cos 2ka,$$

E_R is the resonance energy,
k is the neutron wave number,

Γ is the total width ($\Gamma = \Gamma_n + \Gamma_\gamma + \Gamma_f + \dots$ the sum of the neutron, photon, fission width...),
g is the spin-weighting factor,
a is the effective nuclear radius, and
σ_{pot} is the non-resonant potential scattering cross section.

In a heavy nucleus, at low energy (< 100 eV, say) the total width Γ is frequently less than several hundred meV. This figure sets stringent limits on the necessary resolution of any spectrometer used in studying such states (see also, the estimate of the Doppler broadening of a narrow resonance given later). The transmission $T(E)$ in the region of a narrow resonance changes very rapidly with energy and the instrumental resolution broadens the resonance line shape in such a

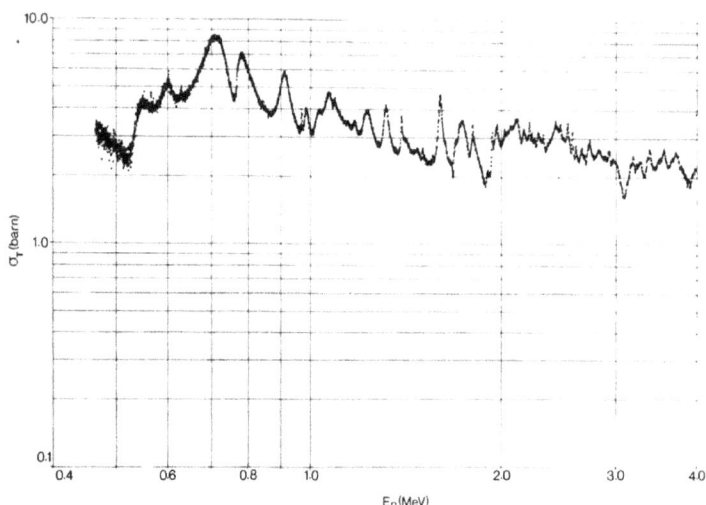

Fig. 1. The observed neutron total cross section of sodium measured at the Karlsruhe cyclotron with a resolution of 0.05 ns/m [13].

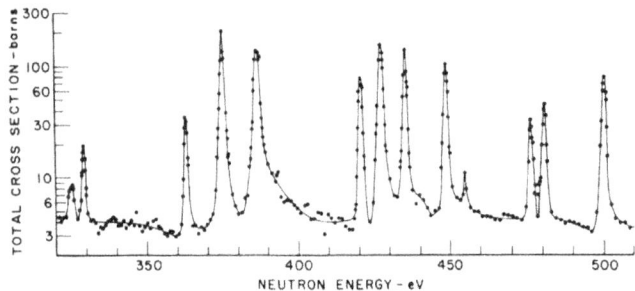

Fig. 2. The observed neutron total cross section of iodine measured at the Nevis Cyclotron of Columbia University with a resolution of 0.5 ns/m [14].

way that the observed transmission $T_{\text{obs}}(E)$ is given by

$$T_{\text{obs}}(E) = \int T(E')\, R(E'-E)\, \mathrm{d}E',$$

where the integral extends over the range of the resolution function $R(E'-E)$.

The factors that contribute to the resolution function in a neutron time-of-flight spectrometer are considered later.

In addition to the effect of resolution broadening of the line shape, there is an intrinsic broadening that must also be taken into account; it is called Doppler broadening, and is due to the thermal motion of the target nuclei. In those cases where this effect is predominant, there is nothing to be gained by indefinite improvements in spectrometer resolution. The Doppler-broadened cross section has the form[16]

$$\sigma_{\Delta}(E') = (\Delta \sqrt{\pi})^{-1} \int_0^\infty \sigma(E'')\exp-\{(E'-E'')/\Delta\}^2\, \mathrm{d}E'',$$

where

$\Delta = 0.32 \sqrt{(E_{\text{R}}/A)}$ eV, the Doppler width.

In a heavy nucleus at an energy of 100 eV, the Doppler width is typically $\Delta \simeq 200$ meV.

If the shape of the resolution function is Gaussian (frequently a good approximation) the observed transmission can be written

$$T_{\text{obs}}(E) = \frac{1}{R(E)\sqrt{\pi}} \int_0^\infty \exp[-n\sigma_{\Delta}(E')]$$
$$\times \exp\left[-\left(\frac{E-E'}{R(E)}\right)^2\right] \mathrm{d}E',$$

where $R(E)$ is the width of the resolution function.

The effect of Doppler broadening on the neutron total cross section for two narrow resonances at 21.9 and 23.5 eV in thorium calculated with parameters obtained initially from "area analyses" of the measured transmission[17] (essentially independent of the two broadening effects discussed above) is shown in fig. 3.

The effect of the finite resolving power of the spectrometer on the Doppler broadened transmission function[17] is shown in fig. 4. In practice, it is necessary to know the shape of the resolution function exactly if the observed function is to be deconvoluted in a reliable way.

Knowledge of the average width $\langle \Gamma \rangle$[18] and the average level spacing $\langle D \rangle$[19,20] of the resonances of

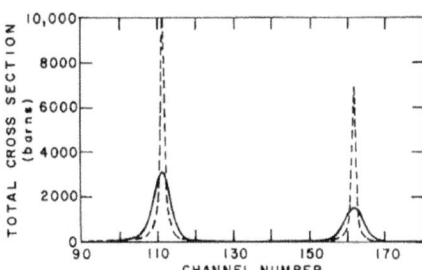

Fig. 3. The effect of Doppler broadening on the neutron total cross section of two sharp resonances in thorium at 21.9 and 23.5 eV [17]). The solid line shows the broadened resonance forms.

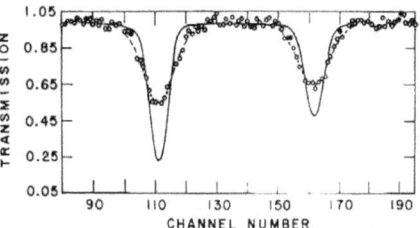

Fig. 4. The effect of the finite resolving power of the spectrometer on the Doppler-broadened transmission function derived from the cross section in fig. 3 [17]).

a given spin and parity, in a particular nucleus, can be used to advantage in setting limits on the spectrometer resolution needed to determine the parameters of individual resonances. We note that, according to Porter and Thomas[18], the most probable neutron width is zero!

In light nuclei ($A \lesssim 20$), neutron induced resonances are observed in the MeV-region. Both the values of their average widths and average spacings are greater than those found at lower energies in medium and heavy mass nuclei[21]. Examples of neutron induced resonances observed in the total cross sections of carbon and oxygen[22] are shown in figs. 5 and 6. The resolving power requirements for studies of the resonant structure of light nuclei are particularly difficult to meet. For example, in the reaction $^{12}\text{C}+\text{n}$ a sharp resonance is observed at a neutron energy of 2.08 MeV[23]. The total width of the resonance is 5 keV and its peak total (resonance) cross section is about 5 b. The resolution needed to observe the *shape* of this resonance, directly, is at least 1 part in 1000.

It is important also to note the decrease in the peak total cross section of s-wave resonances as the neutron energy increases. At $E_n = 100$ eV,

154

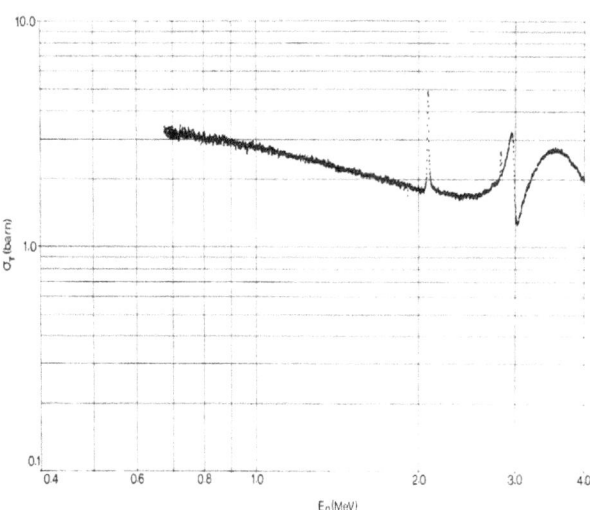

Fig. 5. The observed neutron total cross section of carbon measured at the Karlsruhe cyclotron with a resolution of 0.05 ns/m [22]).

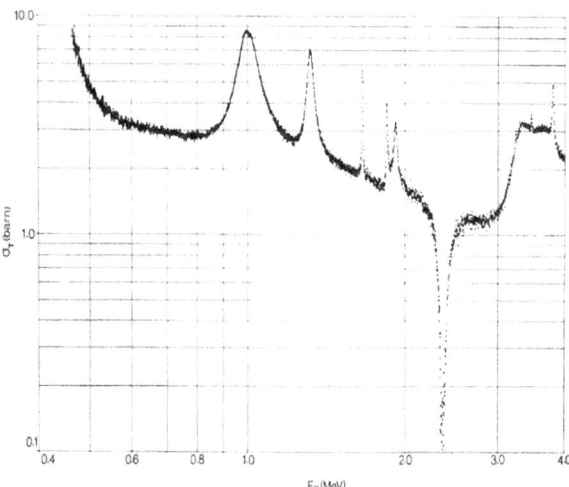

Fig. 6. The observed neutron total cross section of oxygen measured at the Karlsruhe cyclotron with a resolution of 0.05 ns/m [22]).

$\sigma_{\mathrm{Res}}^{l=0} \simeq 4\pi\lambda_{\mathrm{Res}}^2 \simeq 26\,000$ b and, therefore, at $E_n = 1$ MeV, $\sigma_{\mathrm{Res}}^{l=0} \simeq 2.6$ b (here, λ_{Res}^2 is the reduced De Broglie wavelength of the neutron at resonance). This effect puts a premium on the overall sensitivity of a neutron spectrometer required for such studies at high energies.

We are now in a position to discuss the critical question of the resolving power of neutron time-of-flight spectrometers.

2.4. RESOLVING POWER

The measurements of the neutron time-of-flight, t_n, and the length of the flight path, l, are uncertain for a number of reasons. The instant at which a neutron is produced in the source is not known within the interval Δt_B, the duration of the neutron pulse, and the instant it is detected is not known within the interval Δt_D. If the source is surrounded by a moderator, the variation in the slowing-down time of neutrons in the moderating material introduces an uncertainty Δt_M. Timing uncertainties of an electronic nature cannot be eliminated entirely; they are associated with the determination of the exact time at which the burst of neutrons is generated, and with the exact time of neutron detection. The use of "zero-crossing" voltage discriminators has made it possible to reduce the variation in time of detection of voltage pulses from scintillation detectors to less that 1 ns for a dynamic range of 100-to-1 in pulse amplitude. The master clock used to measure the time-of-flight t_n must be stable over adequately long periods, and must be calibrated against a suitable standard.

The length of the flight path, l, is uncertain because the neutron producing target and the neutron detector necessarily have finite thicknesses. In general, neutrons are produced throughout the volume of the primary target, and they are detected throughout the volume of the detector. The neutron detection efficiency is a function of the number of detecting nuclei per unit area, and the neutron cross section, which depends on neutron energy.

If the timing uncertainty from all sources is Δt_n, and the distance uncertainty from all sources is Δl then, for a relativistic neutron of rest energy E_{on}, the uncertainty ΔE_n in measuring the kinetic energy E_n is

$$\Delta E_n = \frac{\beta_n^2}{1-\beta_n^2} (E_n+E_{0n}) \left[\left(\frac{\Delta l}{l}\right)^2 + \left(\frac{\Delta t_n}{t_n}\right)^2\right]^{\frac{1}{2}},$$

where $\beta_n = v_n/c = l/t_n c$.

At non-relativistic speeds, this expression reduces to the familiar form

$$\Delta E_n^{\text{N.R.}} = 2 E_n^{\text{N.R.}} \left[\left(\frac{\Delta l}{l}\right)^2 + \left(\frac{\Delta t_n}{t_n}\right)^2\right]^{\frac{1}{2}}.$$

If, as is often the case, $\Delta l/l \ll \Delta t_n/t_n$ then

$$\left(\frac{\Delta E_n}{E_n}\right)^{\text{N.R.}} \simeq 2\frac{\Delta t_n}{t_n} = 2v_n\frac{\Delta t_n}{l}.$$

The quantity $\Delta t_n/l$, given in ns/m, is therefore useful for comparing the resolution of different time-of-flight spectrometers. We shall see that, at the present time, a resolution of 0.1 ns/m at 1 MeV is considered to be good.

The resolving power of a spectrometer is closely connected with the questions of neutron flux and of signal-to-background. It is not enough simply to know the total flux at the neutron source; the ability to carry out an experiment depends, in an essential way, on the neutron energy spectrum of the pulsed source, and on the flux arriving at the detector. These questions are considered in section 4.

3. Pulsed neutron sources

3.1. GENERATION OF PULSED BEAMS OF IONS

The two main methods used to produce pulses of ions are beam chopping and beam bunching. The deflection of a beam of charged particles by an electric field established between a pair of deflector plates has been used in physics since the work of J. J. Thomson on the deflection of «cathode rays» in 1897. The theory of the deflection of positive ions across an aperture to form well-defined pulses was given by Turner and Bloom[24] and an account of the method is given by Neiler and Good[25] in their definitive article on time-of-flight techniques. The details of the method are discussed in the above references and will therefore not be reproduced here.

It is necessary to produce pulses of ions with:
a) short durations, typically in the ns range,
b) high peak currents, typically greater than several mA,
c) high repetition rate, typically up to several MHz,
d) low current between pulses,
e) small energy spread,
f) small angular divergence.

It is now common practice to combine beam chopping with some form of beam bunching.

3.2. TIME COMPRESSION METHODS

3.2.1. *Klystron bunching*

A beam of charged particles can be velocity-modulated to produce pulses of short duration and high current. This method was first applied to the bunching of electron beams in the development of powerful klystron amplifiers operating in the GHz frequency range[26]. Beams of protons and deuterons were subsequently velocity-modulated to produce pulses of neutrons via (p, n) and (d, n) reactions with durations as short as 1 ns.

If we wish to time-compress a pulse of monoenergetic charged particles moving in a straight line with constant speed $v(0)$, it is necessary to change the speed of a particle, which follows the first particle in the pulse by a time t, to a new value $v(t)$ such that

$$v(t) = v(0)/(1-t/T),$$

where T is the time for the first (unmodulated) particle in the pulse to travel a distance L along a field-free drift tube $[T = L/v(0)]$. The necessary speed $v(t)$ can be achieved by applying an accelerating voltage $V(t)$ across a gap at the entrance to the drift tube. For non-relativistic particles, the optimum voltage $V(t)$ has a quadratic dependence on the time t; in practice, however, $V(t)$ is usually a sinusoidal function of time. The modulating voltage is impressed upon the unmodulated beam, and introduces an irreversible energy spread that may be unacceptable from the point-of-view of the properties of the subsequent particle accelerator, or from the resolution requirements of those experiments where the particle beam is to be used directly for reaction studies.

In 1955, Ashby et al.[27] successfully used klystron bunching of a pulsed beam of 500 keV deuterons from a Cockcroft–Walton accelerator to produce pulses 3 ns in duration from initial pulses 20 ns in duration. Flerov and Tamonov[28] bunched 35 ns wide pulses of 200 keV deutrons by a factor of 70 to give pulses 0.5 ns in duration each with a peak current of 5 mA. However, Delaney[29] had only limited success in klystron bunching 5 keV protons; space-charge debunching effects clearly effected the performance of this system at such a low ion beam energy. (In high-powered electron klystrons, the electron energy is usually greater than 200 keV.)

Positive ion sources with klystron bunching are now available commercially[30]. Peak pulse currents

in excess of 10 mA (protons) with durations of 1 ns are readily achieved on target.

3.2.2. Mobley-magnet bunching

By 1960, another method of time-compression of a beam of charged particles had become practicable; it was based on a proposal made by Mobley in 1952[31]. The basic features of Mobley's method are shown in fig. 7. A pulse of protons with an initial duration Δt_0, enters electrostatic deflection plates to which a time-varying voltage is applied; the pulse of ions emerging from the plates is dispersed linearly as shown. Particles in the dispersed pulse enter a magnetic field and travel along different radii of curvature. The first particles in the original pulse arrive at the focal point having taken a certain time to travel their particular path. The last particles in the pulse travel along a shorter route and arrive at the focus at the same time as those particles that originated at the beginning of the initial pulse. By careful design of the electrostatic deflection system and the magnetic field, pulses of protons have been compressed by a factor of 10. (1 mA peak current, 10 ns duration → 10 mA, 1 ns.)

It should be noted that the introduction of angular divergence into the final bunched pulse may be too large for certain subsequent experiments.

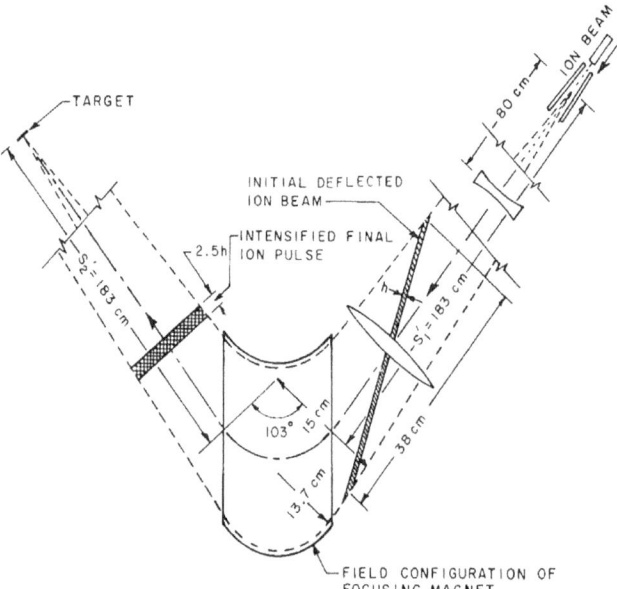

Fig. 7. A schematic diagram of the Mobley magnet time-compression system used on pulsed Van de Graaff accelerators[31].

Several highly successful Mobley bunching systems have, however, been in operation for more than a decade; they are particularly well-suited to low-energy Van de Graaff accelerators[32]).

3.2.3. *Phase bunching*

Bohm and Foldy[33]) and Allwood[34]) showed that the initial motion of the ions in any type of cyclotron is such that phase-bunching occurs *before* entry into the dees. Cohen[35]) solved the equations of motion for an ion in the source region of a fixed frequency cyclotron, and obtained expressions for the phase width $\Delta\theta$ in two extreme circumstances. In the first, it was assumed that the ions leave the source at time zero with zero energy, in which case

$$\Delta\theta \simeq \frac{eV_0}{2md^2\omega^2},$$

and, in the second, it was assumed that the ions instantly gain the maximum energy from the full dee-potential, in which case

$$\Delta\theta \simeq 1.32\left(\frac{eV_0}{2md^2\omega^2}\right)^{\frac{1}{2}},$$

where m and e are the mass and charge of the ion, V_0 is the maximum dee-voltage, d is the dee-spacing, and ω is the cyclotron radial frequency (all in cgs units). Measurements show that the actual values are somewhere between the above extremes[78]); values of $\Delta\theta$ are typically in the range 0.15–0.35 rad, corresponding to pulses with durations of 1–5 ns.

3.3. MECHANICAL ROTORS

Although some improvements in the original mechanical rotor experiment of Dunning et al.[1]) were reported[36]), it was not until copious fluxes of low-energy neutrons became available at nuclear reactors that the field of neutron time-of-flight spectroscopy using mechanical neutron choppers came of age.

The first successful program was (appropriately) carried out by Fermi and his collaborators[2]) in the early 1940's. The first "Fermi chopper" consisted of a sandwhich of aluminum and cadmium foils contained in a steel cylinder, less that 2″ in diameter, that could be spun at 15 000 rpm. The spinning system was viewed with light beams and photocells to give a direct measure of the rotation rate, and a sharp, synchronized pulse from which the time-of-flight of the transmitted neutrons could be measured.

Brill and Lichtenberger[37]), using an improved version of a Fermi rotor, carried out measurements of the neutron total cross sections of several nuclei at energies up to 1 eV. The system enabled six energy intervals to be studied simultaneously. The duration of the neutron pulse was 40 μs. Shortly afterwards, Selove[38]) developed a "fast chopper" ("fast" referring to the speed of the transmitted neutrons), consisting of a steel cylinder with slots cut along the periphery, parallel to its axis. The cylinder rotated about its axis at speeds up to 15 000 rpm, and generated neutron pulses of several μs duration. The considerable length of the steel cylinder provided good shield-

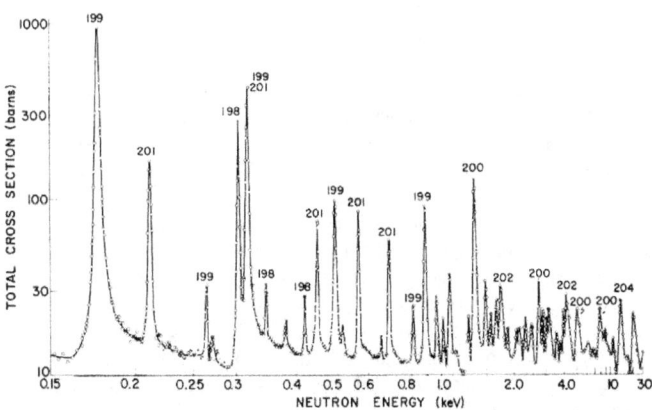

Fig. 8. The observed neutron total cross section of mercury measured at the Argonne fast chopper facility with a resolution of 12 ns/m [40]).

X. TIME-OF-FLIGHT SPECTROMETERS

ing from the higher energy neutrons and gamma-rays from the reactor, and permitted measurements to be made at energies up to 100 eV. By the late 1950's, neutron choppers were used to cover the energy range from 10^{-4} eV to about 10^4 eV. The design of these choppers varied considerably, depending on the required size of neutron beam and range of neutron speed to be passed. Fermi's "slow" chopper was designed for experiments at energies below about 1 eV. It had a symmetrical slot so that the neutron pulse was formed at the center of the rotor (the slot chopped at both ends). The "fast" chopper, built by Selove, contained a divergent slot which chopped only at the entrance of the neutron path. In a symmetrical slot design, the pulse width is proportional to the angular velocity of the rotor whereas, in a divergent slot design, the pulse width is proportional to the tip speed of the rotor. The maximum attainable speed of rotation is, of course, limited by the strength of materials: using steel, or nickel alloys, or fiber glass, the limiting tip speed of a practical rotor is about 10^5 cm/s[39]. Using such a device with narrow slots it is possible to reduce the pulse width to about 0.5 μs and retain a useful flux of neutrons at the detector position (flight paths ranging in length from several meters to several hundred meters have been used)[40]. The resolutions achieved with such devices are adequate for neutron studies up to about 10 keV (see fig. 8), but at higher neutron energies they are no longer competitive with time-of-flight spectrometers using high current particle accelerators capable of producing intense neutron pulses in the ns range.

3.4. CYCLOTRONS

In 1941, Baker and Bacher[5] successfully modulated the arc source of the original cyclotron at Cornell, and produced pulses of 1.4 MeV deuterons with durations between 50 and 100 μs, at repetition rates of 400 pulses per second. Their work was significant because it demonstrated that pulses of energetic charged particles could be obtained by modulating the ion source, a technique that had been tried previously, without success[3].

Shortly afterwards, Rainwater and Havens[41] began their long and fruitful collaboration in neutron time-of-flight spectroscopy at Columbia University. Their work formed the basis of future developments in the field, particularly at neutron energies below several keV. An example of their pioneering work is shown in fig. 9; it represented a measure-

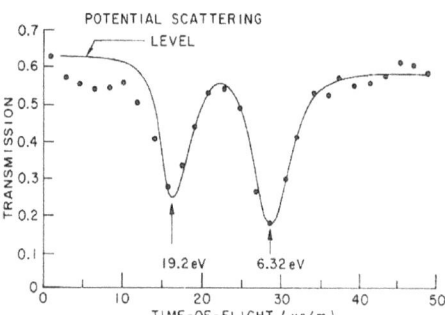

Fig. 9. Early measurements of neutron resonances in antimony obtained using the time-of-flight spectrometer at Columbia University in the 1940's [42].

ment with greatly improved resolution and statistical accuracy compared with previous work. They also developed the "area-method" of analysis of the observed data to obtain Breit–Wigner resonance parameters, independently of the resolving power of the spectrometer[42]. Later, the Columbia University Nevis 340 MeV Synchrocyclotron was modified for use in low-energy neutron time-of-flight spectroscopy by Rainwater et al.[43].

In this system, a pair of deflecting plates was placed at a mean orbit radius of 69" (the maximum useful radius being 73") and used to deflect the proton bunch onto a 0.75" thick lead plate below the median plane. At 300 MeV, the proton range in lead is a few inches and is comparable with the mean free path for a nuclear collision. At least one neutron is produced for every incident proton: the energy spectrum of the neutrons consists of two components, the first is a characteristic evaporation spectrum which is isotropic, and extends up to about 10 MeV, and the second is a direct interaction spectrum which is strongly forward-peaked and contains neutrons with energies up to the maximum allowed by kinematics. The duration of the entire deflected proton bunch was about 20 ns. The primary Pb(p, n) spectrum was moderated using polyethylene slabs and the resulting neutron spectrum passed along a helium-filled 200 m flight path. It was necessary to place a thick filter of lead in the beam to attenuate the intense flux of photons that accompanied the reaction. An example of the high resolution achieved with this spectrometer is shown in fig. 10[44].

For more than two decades, the 150 MeV proton synchrocyclotron at Harwell, England has been used for neutron time-of-flight studies[45]. Proton pulses, a few nanoseconds in duration, are produced

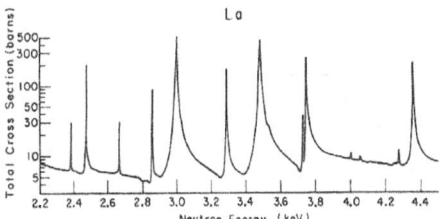

Fig. 10. The observed neutron total cross section of lanthanum in the keV-region measured at the Nevis cyclotron with a resolution of 0.5 ns/m [44]).

by a beam deflection system, similar in principle to that used at the Nevis cyclotron. The early work with this facility concentrated on measurements of neutron total cross sections between 10 and 140 MeV, and on studies of polarization effects in (p, n) reactions, also at high energies. In 1965, the system was developed to cover the neutron energy range down to 250 keV [46]), and in 1968, the accelerator was improved to provide higher peak currents and a higher repetition rate (800 Hz). Additional flight paths were also added so that many simultaneous measurements became possible at energies in the eV-region [47]). It is interesting to note that the first observation of Mott–Schwinger scattering of fast neutrons by a heavy nucleus [48]), and the first use of a solenoidal field to precess the magnetic moment of a neutron in polarization studies [49]) were made with this accelerator.

In recent years, the Karlsruhe sector-focused cyclotron has been used with success to study neutron interactions in the MeV region [50]). The machine produces high intensity pulses of 50 MeV deuterons with narrow pulse widths, typically less than 2 ns. Normally, the microstructure pulses occur at repetition rates between 10 MHz and 30 MHz which are too high for time-of-flight studies of neutrons with energies below several MeV (due to the overlap of the slower neutrons from one cycle with the faster neutrons of the next cycle). The sacrifice in intensity that comes about from a straightforward reduction of the pulse repetition rate, achieved by suppressing most of the microstructure pulses, is too great and therefore it is necessary to reduce the rate and, at the same time, to maintain as much of the high average beam intensity of the cyclotron as possible. These conflicting requirements have been largely met by a combination of beam bunching and beam deflection: the repetition rate has been reduced from 33 MHz to 0.2 MHz while the average beam inten-

sity has decreased by only a factor of three [50]). This factor is a consequence of suppressing two out of the three microstructure pulses normally present in the cyclotron when operating in its 3rd harmonic mode. Examples of the high resolution data obtained with this system are shown in fig. 11 [50]).

The major developments in the generation and use of nanosecond pulses of protons and deuterons from cyclotrons reported during the period between 1955 and 1960 have been discussed in detail by Bloom [78]), and will not be repeated here. However, one significant event that happened during that period should not go unmentioned namely, the first observation of an isobaric analog resonance in the ^{51}V(p, n) reaction by Anderson and Wong [79]), who used the nanosecond neutron time-of-flight spectrometer associated with the cyclotron at the Lawrence Livermore Laboratory in California. Their work led directly to the formation of an important branch of the Nuclear Physics of the 1960's.

The characteristics of several cyclotrons presently in use for neutron time-of-flight studies are listed in table 2.

3.5. ELECTRON LINEAR ACCELERATORS

Following the original analysis of the properties of disc-loaded waveguides by Cutler [51]), travelling-wave electron linear accelerators were developed by Fry et al. [52]) at the Telecommunications Research Establishment at Malvern, England and, independently, by Ginzton et al. [53]) at Stanford University in

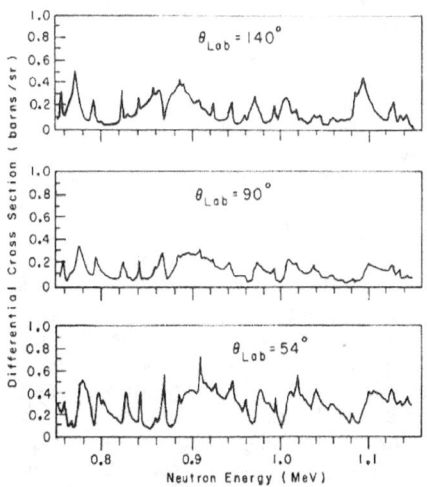

Fig. 11. The observed differential cross section of calcium measured at the Karlsruhe cyclotron [50]).

160

550   F. W. K. FIRK

TABLE 2

Several cyclotrons presently in use for neutron time-of-flight studies.

Facility	Energy (MeV)	Particle	Pulse duration (ns)	Repetition rate (Hz)	Neutrons/s (peak)	Neutrons/s (average)
Karlsruhe isoch. cyclotron	50	d	1.5	2×10^4	7×10^{18}	2×10^{14}
				2×10^5	7×10^{17}	2×10^{14}
Kiev isoch. cyclotron	60	d	1	2×10^4	1.7×10^{18}	3.4×10^{14}
Harwell synchrocyclotron	150	p	4	8×10^2	3×10^{19}	1×10^{14}

the U.S.A. The first successful operation of a 1 m long accelerator was reported by Fry et al.[52]); it was powered by a magnetron operating at 3000 MHz, and produced electron pulses 2 μs in duration with peak currents of 35 mA, at a repetition rate of 200 Hz. The energy was 0.5 MeV. In the same year, the Stanford University group reported their work and achieved essentially the same energy gain per unit length as the Malvern work. The general design parameters of these two pioneering developments were remarkably similar and, indeed, very little has changed in the design of such systems in the intervening years.

Whilst the Stanford group concentrated their efforts on the design and construction of high-energy electron accelerators, with a view to the study of nuclear structure by means of electron scattering, the development in England concentrated on the building of low-energy, high current accelerators suitable for the production of intense pulses of neutrons via the photo-disintegration of suitable nuclei[9]) (beryllium for electron energies below 5 MeV and natural uranium for energies above 15 MeV). Two other developments in the U.S.A., one at Yale University[54]) and the other at MIT[55]), were also underway in the late 1940's using standing-wave electron linear accelerators. These accelerators were used in early studies of neutron- and photon-induced reactions, and were important in establishing experimental techniques that proved to be of lasting value. For example, at Yale some of the first measurements of the spectra of γ-rays from low-energy neutron resonances were carried out[56]), and at MIT, emphasis was placed on the production of very short bursts (several ns in duration), and the study of photo-neutron spectra in the energy range from several hundred keV to several MeV[57]). These standing-wave accelerators were limited in the peak currents available, and were eventually

replaced by high-powered travelling-wave machines.

Large scale neutron time-of-flight facilities, using electron linear accelerators to provide intense fluxes of photo-neutrons were introduced at Harwell, England between 1952 and 1959. In June 1952, a 15 MeV travelling wave electron linear accelerator, designed and constructed by Mullard Ltd.[58]), was installed at Harwell replacing Fry's 3.5 MeV accelerator which had been used for neutron time-of-flight studies for a period of three years. The 15 MeV accelerator produced pulses of electrons, about 1 μs in duration, to bombard a 2″ diam. × 2″ long cylinder of natural uranium, surrounded by 1″ thick slabs of moderating material (lucite). The peak pulse current was 25 mA and the maximum repetion rate 400 Hz. The neutron producing target was viewed, initially, by three flight paths, one for the study of the elastic scattering of neutrons, using an annulus of $^{10}BF_3$ counters as detectors[59]), another for the study of γ-rays resulting from neutron capture[60]) and the third for neutron total cross section measurements[61]). The first complete set of neutron resonance parameters was obtained for the low-energy resonances of the silver isotopes, using this system[59]). Two other developments took place there between 1955 and 1958. In the first, Poole et al.[62]) installed a pulsed magnet at the output of the accelerator so that the electrons could be deflected on a pulse-to-pulse basis into different target areas at any desired repetition rate, provided that the total rate did not exceed 400 Hz. In one of the beam-switched modes, a low repetition rate (10 Hz) beam of electrons struck a uranium target at the center of a large reactor-type lattice, to provide a pulsed source for time-of-flight studies of the spectra of very low-energy neutrons from different lattice arrangements. The majority of the electron pulses struck the standard target and permitted the standard experiments to continue. In the second development[63]), the electron gun was pulsed from a separate modulator and the pulse width reduced from greater than 1 μs to less than 200 ns. The neutron flux only decreased by a factor of two in this mode, due to an increase in the electron energy as a result of less beamloading and as a consequence of injection at the very beginning of the rf pulse (the so-called "stored-energy" mode). Using a flight path 57 m long, a nominal resolution between 2 and 3 ns/m was achieved for total cross section measurements in the keV-region[64]).

In 1959, a 28 MeV electron linear accelerator was installed at Harwell with many new features that permitted simultaneous studies of neutrons over an energy range from 10^{-2} eV to 10^7 eV [65]). The main neutron-producing target consisted of a sub-critical assembly of uranium-235 with appropriate control rods and moderator; a gain of greater than ten was achieved, coupled with a relaxation time of about 0.2 μs. Furthermore, the electron beam could be switched into three different target areas, each suitable for completely different experiments [the subcritical assembly, viewed by eight flight paths, ranging in length from 5 to 300 m; an unmoderated target for photo-neutron time-of-flight studies using 5 ns pulses, and flight paths up to 110 m [66]); and reactor-lattice assemblies]. In its most advanced form, this beam-multiplexing system could provide beam-switched pulses ranging in width from 5 ns to 5 μs, with energies between 5 MeV and 28 MeV, all on a pulse-to-pulse basis[66]).

Another large neutron time-of-flight facility was put into regular operation by 1960 at Saclay, France. The accelerator was similar in performance to the Harwell machine[67]). A full range of neutron total, capture, scattering and fission cross sections were measured: a comprehensive computer-based data acquisition system was used for these measurements[68]).

It was not until 1961, however, that the first high-powered electron linear accelerator representing a new generation of machines, was commissioned at Yale University[69]). This machine produced pulses 0.1 μs in duration with peak currents of 1.2 A at a rate of 700 Hz. The beam power on the target in the long (5 μs) pulse mode was 32 kW (at an electron energy of 40 MeV). In 1966, this system was improved by installing a ns modulator for the triode electron gun[70]) and by 1968, pulses of electrons 10 ns wide and 10 A in peak current, with energies up to 70 MeV were obtained on a regular

basis. Accelerators similar in design to the Yale machine were installed at San Diego and at R.P.I.[71]). All these machines had extensive time-of-flight facilities capable of covering the neutron energy from thermal energies up to 50 MeV [72]).

The most advanced system of this generation of accelerators has been in operation at the Oak Ridge National Laboratory since the early 1970's. Very high resolution studies have been carried out with this facility (called ORELA) in the areas of neutron total, capture, scattering and fission cross sections[73]). Neutron inelastic scattering measurements and studies of the spectra of γ-rays from resonances have also been made. The neutron beam has been polarized at low energies by transmission through a target of dynamically polarized protons[74]), and the polarized neutrons used in studies of the spin-dependence of resonances, particularly in fission, at energies up to several keV (fig. 12)[75]).

The characteristics of several high-powered electron linear accelerators, either in use or under construction, are listed in table 3.

Picosecond pulses from electron linear accelerators.

Norris and Hanst[76]), working at the linear electron accelerator laboratory of the Edgerton, Germeshausen and Grier corporation in 1965, first developed electron pulses of very high current and very short duration. They made significant technical advances including the isolation of a single rf pulse of the accelerator (operating at 1300 MHz), the synchronization of the entire system to the fundamental frequency, the development of stable high-powered pulse modulators to trigger high-current

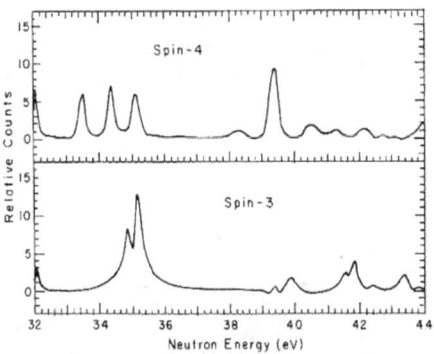

Fig. 12. The clear-cut determination of the spins of resonances in the fission cross section of ^{235}U achieved by using both polarized incident neutrons and polarized target nuclei[75]). The measurement was made at the Oak Ridge Electron Linear Accelerator.

F. W. K. FIRK

Survey of some electron linear accelerators in use or under construction.

Facility	Pulse duration	Repetition rate (Hz)	Neutrons/s (peak)	Neutrons/s (average)
Oak Ridge, U.S.A. (ORELA) (1970–) 140 MeV LINAC	3 ns 1 μs	1000 700	3×10^{18} 1×10^{17}	1×10^{13} 7×10^{13}
Kurchatov, U.S.S.R. 60 MeV LINAC	50 ns 5 μs	50 150	3×10^{17} 3×10^{17}	1×10^{12} 2.5×10^{14}
Geel, Belgium Improved (1979–) 120 MeV LINAC	3 ns 2 μs	900 900	4×10^{18} 2.5×10^{16}	1×10^{13} 4.5×10^{14}
Harwell, U.K. 130 MeV LINAC (1979-)	0.1 μs 5 μs	300	3×10^{18} (with booster) 2×10^{17}	1×10^{14} 3×10^{14}

coaxial triode electron guns, the use of subharmonic bunching of the electrons that resulted in an increase in the peak current of a factor of 75, and the development of beam monitoring methods to provide relatively undistorted measures of the final current pulses. By 1967, current pulses lasting 50 ps, and containing 10^{10} electrons per pulse, with energies in the range 5–20 MeV, were available.

A system, based upon that of Norris and Hanst, has recently been installed at the Argonne National Laboratory[77]), and is being used for measurements of the energy spectra of photoneutrons emitted from states in nuclei at excitation energies below 20 MeV.

The pulse durations are about 30 ps and the peak electron currents are 200 A. The excellent resolution achieved with this system is illustrated in fig.

13. In this example, the resolution is limited by time-response of the neutron detectors.

3.6. VAN DE GRAAFF ACCELERATORS

The continuous current from a Van de Graaff accelerator can be readily changed into a series of pulses of short duration by using one or more of the methods outlined above. From the beginning, it was recognized that it is generally advantageous to generate the pulses before, rather than after, acceleration; less power is required to modulate a low-energy ion beam, increased peak currents can be generated by klystron bunching of a low-velocity ion beam, and beam deflection across slits at the terminal creates low-energy background radiation far from the final experimental area of the accelerator. The technical problems of terminal pulsing, however, are usually more severe than those associated with beam deflection at the output of the accelerator therefore, the majority of the early Van de Graaff accelerators used the method of post-deflection of the beam[25]).

By 1960, more than a dozen Van de Graaff accelerators were using some form of ns pulsing for neutron time-of-flight studies[78]). Particularly interesting programs were underway at Los Alamos[80]) and Oak Ridge[7]) in the U.S.A. and at Aldermaston[81]) in England. The method of combining a neutron source which produced a narrow energy spectrum from a suitable (p, n) or (d, n) reaction, with the ns time-of-flight method enabled a clear separation to be made between elastically and inelastically scattered neutron groups for the first time (see fig. 14)[80]).

One of the programs at Oak Ridge, using a pulsed 3 MV Van de Graaff accelerator concentrated on neutron studies in the region between 2 keV and 100 keV. In this system, the ^{7}Li(p, n)^{7}Be

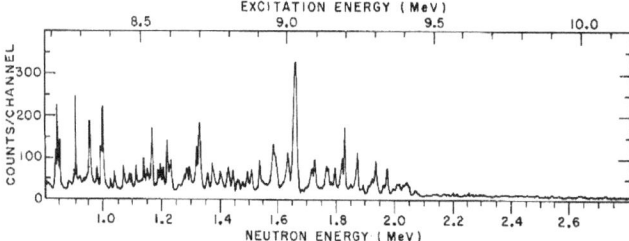

Fig. 13. High resolution studies of the ^{208}Pb(γ, n_0) reaction made at the Argonne Electron LINAC using electron pulse widths of about 30 ps with peak currents of 200 A [77]).

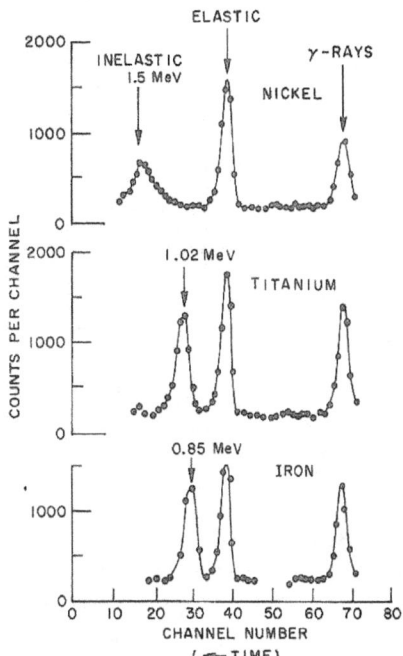

Fig. 14. Some of the earliest measurements of neutron inelastic scattering from several elements using a pulsed Van de Graaff accelerator to generate 3 ns pulses of protons[80]. The incident neutron energy was 2.45 MeV.

reaction was used, just above threshold, to produce neutrons, directly, in the keV-region. The neutrons were emitted in a forward cone as a result of the kinematics of the reaction[82]. The energies available from the reaction, given as a function of the difference between the proton energy and the threshold energy for various angles of emission are shown in fig. 15. The unique features of this low-energy, nearly monoenergetic, pulsed source of neutrons were exploited in a series of pioneering studies of neutron radiative capture[83]) and of the spectra of photons from neutron radiative capture in the keV-region[84]). The systematic study of Macklin and Gibbons[85]) of neutron capture cross sections of very small quantities of separated isotopes, gave us information of critical importance in the field of nucleosynthesis.

In recent years, ns pulsing of ion beams from Van de Graaff accelerators has become a standard tool, providing a method for both neutron and charged particle spectroscopy. Detailed reviews of the pioneering work in the field are given in ref. 86

(note, particularly the articles by Stelson and A. B. Smith).

3.7. COCKCROFT-WALTON GENERATORS

Although relatively little work has been reported in the field of neutron time-of-flight spectroscopy using pulsed Cockcroft–Walton generators, two such programs played an important part in the development of the field in general. In 1940, Fertel et al.[4]) at the Cavendish Laboratory in Cambridge, modulated the ion source of their generator and carried out one of the very first accelerator-based time-of-flight experiments and, in 1955, Ashby et al.[27]) achieved a notable breakthrough in technique by Klystron-bunching a beam of 500 keV deuterons from a Cockcroft–Walton generator at Livermore, California, to give 3 ns pulses at a high repetition rate.

3.8. NON-STANDARD METHODS

3.8.1. Pulsed reactors

About thirty years ago, Frisch proposed building a pulsed neutron source at Harwell, England using two sub-critical assemblies, each mounted on a

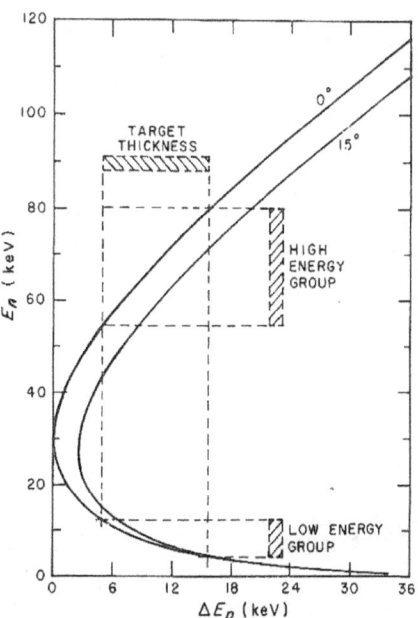

Fig. 15. Kinematics of the ^{7}Li(p, n)^{7}Be reaction close to threshold[82]).

X. TIME-OF-FLIGHT SPECTROMETERS

mechanical rotor so that they could be brought close together for a period of a few microseconds, at a reasonably high repetition rate. The system was considered to be technically difficult to construct, and not without its risks. It was never built. By 1960, however, a pulsed reactor had been built at Dubna in the USSR, and was subsequently used for more than a decade for neutron spectroscopy in the energy range from a few MeV to several keV. In its earliest form, the pulsed reactor operated at a thermal power level of only 3 kW, and was found to be equivalent to a "stationary" (!) reactor with a power level of almost 30 MW [87]. The small average power level resulted in a very low ambient background and permitted the study of processes with very small cross sections. For example, the first measurements of (n, α) resonant cross sections (fig. 16) were made with this system, together with some of the earliest experiments using ultra-cold neutrons[88]. In 1965, the facility was improved by operating the pulsed reactor in synchronism with a pulsed electron linear accelerator so that the reactor acted as a dynamic multiplier of neutrons emitted from the target of the accelerator. In contrast to the static booster used at the Harwell electron linear accelerator[65], with its gain of about 10, the multiplication factor of the pulsed booster at Dubna reached values of several hundreds. In this mode, the pulse duration was typically 5 μs. Although this system did not compete in energy resolution with the newer pulsed electron linear accelerators and cylotrons, it was a prolific source in the low-energy region where adequate resolution could be achieved, and where large, effective detectors with long response times could be used [as in the (n, α) studies]. In 1970, the system was further improved to include a 25 kW pulsed reactor and a new high

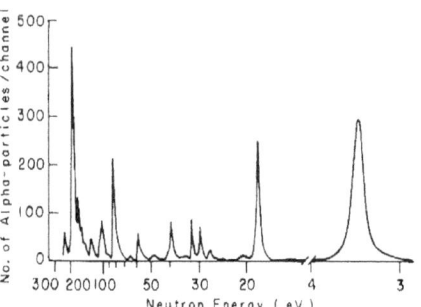

Fig. 16. The observed relative yield of alpha particles from low-energy resonances in samarium obtained with the Dubna pulsed reactor[87].

TABLE 4

Design parameters of the Dubna pulsed reactor–electron linac combination.

Parameters of the IBR-2 reactor	
Mean thermal power	4 MW
Width of power pulse	90–100 μs
Pulse repetition rate	5–50 Hz
Power during pulse:	
a) at 5 Hz	7700 MW
b) at 50 Hz	770 MW
Power released between pulses	0.22 MW
Neutron generation time	4×10^{-8} s
Fuel loading	90 kg

Parameters of LIU-30 electron accelerator	
Electron energy	30 MeV
Pulse current	250 A
Pulse width	0.5 μs
Repetition rate	50 Hz
Beam power (average)	0.2 MW
Neutrons per pulse from thick natural uranium target	1.2×10^{13} per pulse

current electron accelerator[87]. Shortly afterwards, the design of a new pulsed reactor with a power level of several MW, coupled to an induction electron linear accelerator was started. The impressive design parameters of this combination are given in table 4 [87].

The expected value of the neutron flux during the power pulse in the center of the active zone is 1.2×10^{18} neutrons/cm$^2 \cdot$s (pulse width 90 μs). In the booster mode, the yield is expected to be 2×10^{16} neutrons/cm$^2 \cdot$s during a pulse 3 μs in duration.

3.8.2. The LAMPF neutron time-of-flight facility

The Los Alamos Meson Physics Facility provides protons with energies up to 800 MeV, in 500 μs pulses at a repetition rate of 120 Hz. The peak current is greater than 15 mA. The general modifications necessary for producing shorter pulses for time-of-flight experiments are shown in fig. 17[89]. Each pulse is modulated at the source to give a short pulse ($<10 \mu$s duration) that is deflected with a pulsed magnet onto a thick target of natural uranium. In this mode, about 1% of the total beam is used for generating neutrons. A proton storage ring is planned so that the 10 μs pulses can be stored and bunched to give pulses several ns in duration. Typical expected perfor-

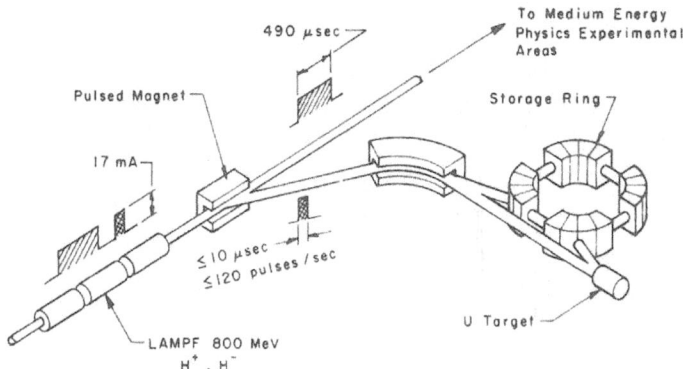

Fig. 17. A schematic diagram of the proposed neutron time-of-flight facility based on the Los Alamos 800 MeV proton linear accelerator[89]). Recently, a pulsed beam was deflected onto a neutron-producing target, but the storage ring has yet to be built.

mance figures of the system are listed in table 5. It is seen that the anticipated fluxes compare favorably with those anticipated from the new high-current cyclotrons and electron linear accelerators. The full potential of this system has yet to be realized.

4. Neutron source characteristics

4.1. Unmoderated sources

The primary energy spectrum of neutrons obtained by bombarding a suitable target with particles from a pulsed accelerator depends upon the nature of the particles and their kinetic energy, and on the target material. The spectrum depends also on the thickness of the target, and in some cases, on the angle of observation of the outgoing neutrons. The interactions most often used to produce neutrons are:

$^2H(d, n)$, $^2H(\gamma, n)$, $^3H(p, n)$, $^3H(d, n)$,
$^7Li(p, n)$, $^9Be(p, n)$, $^9Be(d, n)$, $^9Be(\alpha, n)$,
$^{12}C(d, n)$, $^{12}C(\gamma, n)$, $^{13}C(\alpha, n)$, and in the naturally occurring heavy elements:

$Ta(\gamma, n)$, $W(p, n)$, $W(\gamma, n)$, $Pb(p, n)$,
$Pb(\gamma, n)$, $Bi(p, n)$, $U(p, n)$, $U(\gamma, n)$, and $U(\gamma, fn)$.

The energy characteristics of some of these reactions, in light nuclei, are listed in table 6.

Examples of the wide range of spectrum shapes produced in different reactions are shown in figs. 18, 19, and 20. In these figures, we see the kinematically collimated, nearly monochromatic spectrum of neutrons from the $^7Li(p, n)$ reaction just above threshold[82]), the characteristic Maxwellian spectrum from the $^{235}U(\gamma, fn)$ reaction[90]), and the spectrum from the $^{nat}U(p, n)$ reaction for incident protons with an energy of 150 MeV, measured in the forward direction[91]). As shown in fig. 19, the spectrum of neutrons from a photon induced reaction in a heavy nucleus has a Maxwellian form with relatively few neutrons produced above 10 MeV (even for incident photon energies up to 100 MeV). It is possible, however, to increase the relative number of high energy neutrons from photo-disintegration by using targets with low mass numbers. The observed (unnormalized) neu-

TABLE 5
The LAMPF neutron time-of-flight facility.

	No storage ring	Storage ring (high duty cycle)	Storage ring (low duty cycle)
Pulse width	5 ns–5 μs	5 ns–200 ns	70 ns–200 ns
Repetition rate (Hz)	120	600–120	low (on demand)
Proton intensity in the pulse (s^{-1})	1×10^{17}	2.5×10^{18}	2×10^{20} to
		2.5×10^{19} to (with bunching)	6×10^{20} (with bunching)
Neutron output per 100 ns pulse	3×10^{11}	2×10^{13}	2×10^{15}
Maximum neutron output (s^{-1})	2×10^{15}	2×10^{15}	2×10^{15}
	(5 μs pulses at 120 Hz)	(100 ns pulses at 120 Hz)	(pulsed at 1 Hz)

556   F. W. K. FIRK

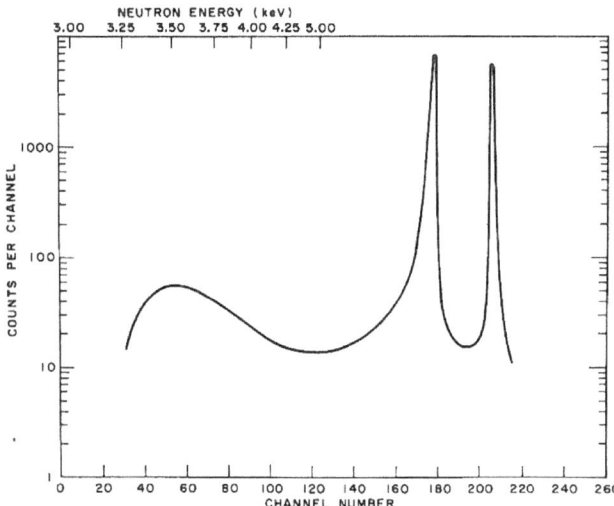

Fig. 18. The observed relative yield of neutrons from the ^{7}Li(p, n)^{7}Be reaction 17.6 keV above threshold[82]. The peak at channel number 210 is due to the detection of γ-rays from the Li target. (Flight path length: 1 m.)

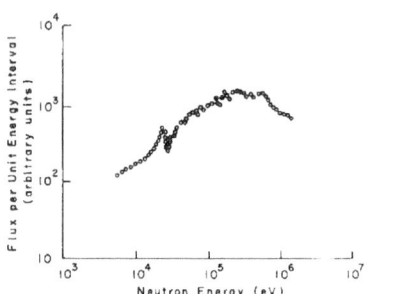

Fig. 19. The shape of the neutron spectrum from the ^{235}U target of the Harwell booster[90].

TABLE 6

Reaction Q-value (MeV)	^{3}H(p, n) −0.764	^{7}Li(p, n) −1.646	^{2}H(d, n) +3.266	^{3}H(d, n) +17.586	^{9}Be(α, n) +5.708
Energy of incident particle (MeV) ↓		Neutron energy at 0° (MeV)			
0	–	–	2.448	14.046	5.266
2	1.201	0.229	5.238	18.259	7.707

tron spectra from the reactions ^{12}C(γ, n), ^{16}O(γ, n) and ^{28}Si(γ, n) obtained when irradiated with 50 MeV bremsstrahlung[92], are shown in fig. 21.

4.2. MODERATED SOURCES

The primary flux of neutrons generated when a pulse of energetic charged particles interacts with a suitable target material has an energy spectrum that contains very few neutrons below a few keV.

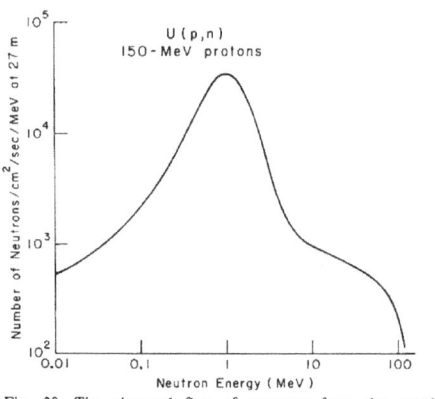

Fig. 20. The observed flux of neutrons from the reaction ^{nat}U(p, n) measured at the Harwell 150 MeV Cyclotron[91].

NEUTRON SPECTROMETERS 557

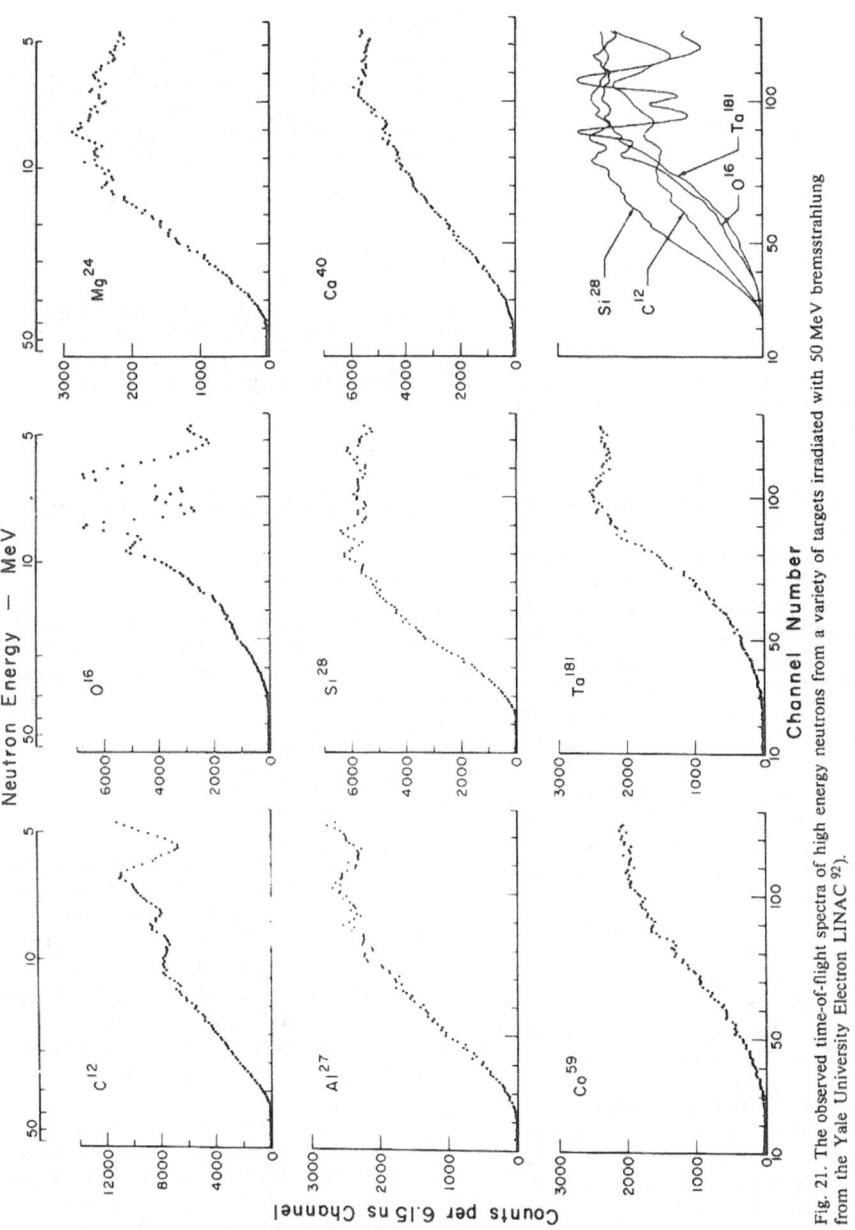

Fig. 21. The observed time-of-flight spectra of high energy neutrons from a variety of targets irradiated with 50 MeV bremsstrahlung from the Yale University Electron LINAC [92].

X. TIME-OF-FLIGHT SPECTROMETERS

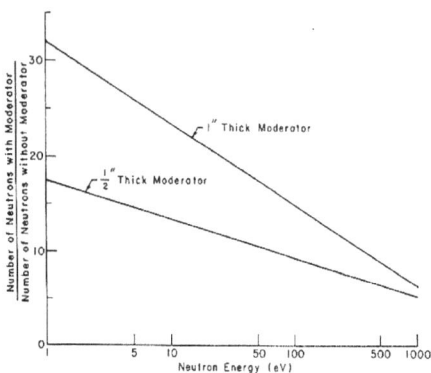

Fig. 22. The effect of moderators of different thicknesses on the yield of low energy neutrons from the reaction ^{nat}U(γ, n) for 15 MeV bremsstrahlung[93].

The flux at low energies can be enhanced considerably by moderating the primary spectrum with an hydrogeneous material. The spectrum at very low energies from a thick moderator has a form that is closely proportional to $1/E_n$ (at typical "slowing-down" type spectrum). The measured effect of a moderator, consisting of a 1″ thick lucite slab, on the photoneutron spectrum from the reactions ^{238}U(γ, n) and ^{238}U(γ, fn) for incident bremsstrahlung with a maximum energy of 15 MeV is shown in fig. 22[93].

If a perfectly sharp pulse of charged particles (or photons) interacts with a target to produce a primary fast neutron flux, the shape of the pulse of neutrons emerging from the moderator is no longer perfectly sharp but has a time distribution that depends on the energy of the emerging neutron. This time dispersion is a consequence of the variations in the time intervals between the multiple collisions suffered by neutrons on passing through the moderator. Ribon and Michaudon[94] studied this effect using a Monte Carlo method to simulate the neutron moderation process in a variety of materials each having a range of thickness. It is possible to obtain a reasonable compromise between the enhancement of the flux at low energies (below a few keV) and the time dispersion of the pulse. For a polyethylene moderator 2 cm thick, the flux at 10 eV increases by a factor of 100 and the time dispersion due to the moderator is 0.5 μs.*

* The time dispersion Δt_m due to a moderator has a $1/\sqrt{E_n}$ dependence on the neutron energy.

4.3. Neutron flux and resolving power

The important questions of neutron flux and resolving power have been discussed in a particularly useful way by Rae and Good[95], as follows:

The number of neutrons N detected per unit time per unit area in an interval ΔE_n at an energy E_n, in a detector placed at a distance l from a pulsed source, is

$$N = I(E_n) \frac{\Delta t'}{4\pi l^2} f \Delta E_n ,$$

where $I(E_n)$ is the instantaneous rate of emission of neutrons from the source, $\Delta t'$ is the duration of the accelerator pulse, and f is the repetition rate. If a constant resolution $R(= \Delta E_n / E_n)$ is chosen, then

$$R = 2[(\Delta l)^2 + (v_n \Delta t)^2]^{\frac{1}{2}}/l,$$

where the symbols have been defined previously. The expression for N can therefore be written

$$N = \frac{I(E_n) \, \Delta t' \, f E_n \Delta E_n R^3}{16\pi \left[(\Delta l)^2 + (v_n \Delta t')^2\right]},$$

in which $\Delta t' = \Delta t$ (the timing uncertainty due to the moderation process is proportional to $1/v_n$, and is equivalent to an uncertainty in the length of the flight path).

The neutron emission rate during the pulse from a moderated source has the form

$$I(E_n) = \frac{\text{(constant) (pulse current)}}{E_n^n \sigma_H^2(E_n)},$$

where $\sigma_H(E_n)$ is the hydrogen scattering cross section, and the constant is a function of the type of bombarding particle, its kinetic energy, the particular reaction used, the target geometry, and the efficiency of the moderator. In practice, the value of n is between about 0.8 and 0.9 (it would be unity for an infinitely thick moderator). This expression is found to describe well the observed shapes of neutron spectra up to energies of about 100 keV.

When using a pulsed, broad-range, neutron spectrum for studies at low energies, it is necessary to avoid the effect of the overlap of the flight-times of neutrons from successive pulses. In these circumstances, it is not possible to use both high repetition rates and long flight paths. However, a thin ^{10}B-filter, with a thickness chosen to transmit a very small fraction of the unwanted neutrons, can be used to improve the situation. The useful flux of neutrons throughout the eV-region is then necessarily reduced and some compromise must be made between filter thickness and accelerator repetion rate.

A comparison between the effective neutron fluxes from different pulsed neutron sources for a constant value of R (arbitrarily chosen to be 8.8×10^{-4}), is shown in fig. 23[95,51]. (These curves take into account the necessary beam-filter, and the effect of moderator dispersion.)

The Karlsruhe cyclotron, with its very high repetition rate, very short pulse width, and high instantaneous deuteron current ($\sim 3\,A$), has the best performance in the MeV-region, the high-powered electron linear accelerator at Oak Ridge (ORELA) has a consistently good performance throughout the keV-region, and the pulsed, sub-critical assembly at Harwell remains most useful at energies below about 1 keV. [This assembly will soon be one target in a new facility based on an electron linear accelerator similar in specification to ORELA[96]), see fig. 24.]

It is clear that the practical production of intense pulsed sources of neutrons from charged particles depends ciritically on the efficiency for converting the original particle energy into a neutron. The generation of neutrons from electrons via the (γ, n) reaction is not very efficient (several GeV of electron energy is required to produce a neutron). The reasons are that the $e \to \gamma$ process is purely electromagnetic, and the (γ, n_{total}) cross sections for heavy nuclei are modest, having appreciable values only in the region of the giant electric dipole resonances (typically between 10 MeV

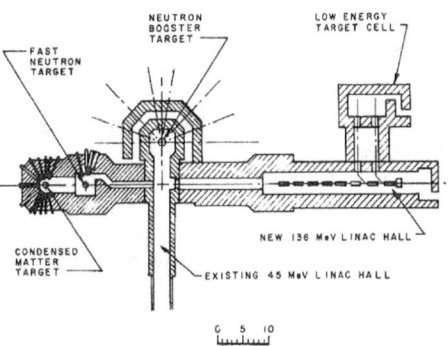

Fig. 24. The layout of the new 136 MeV Electron LINAC at Harwell[110]). Note the use of four independent target areas that permit simultaneous measurements to be made at energies ranging from sub-thermal (condensed matter studies) to about 100 MeV (photoneutron studies).

and 20 MeV). Nonetheless, electron linear accelerators are popular in this field – they are relatively inexpensive, they lend themselves to multiple targets, multiple pulse lengths, and the use of multiple flight paths so that their overall use can be made very high indeed.

The production of neutrons via (p, n) spallation reactions using high-energy protons as projectiles requires only 50 MeV per neutron at an incident energy of 1 GeV. If target-cooling determines the maximum pulsed source intensity then high-energy protons are most attractive as projectiles [see, for example, the exciting but unhappily defunct proposal for an intense neutron generator at Chalk River[97])].

5. Neutron detectors

A detailed discussion of neutron detectors is given by Harvey and Hill and by Grosshoeg in other papers in this volume; only the essentials will therefore be presented in this section. The desirable features of a neutron detector used in a time-of-flight spectrometer are:

1) high efficiency which is a slowly varying function of energy,
2) low efficiency for detecting background radiation,
3) fast time response, preferably in the ns range,
4) ease of manufacture in various shapes and sizes,
5) simplicity of the associated electronics.

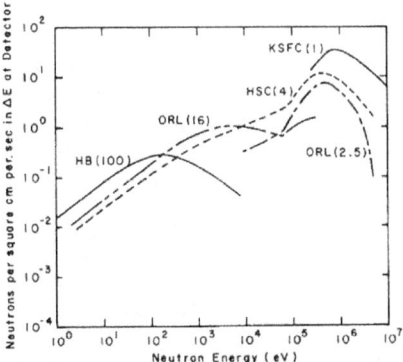

Fig. 23. A comparison of the flux of neutrons from different pulsed sources for a fixed resolution of $\Delta E/E = 8.8 \times 10^{-4}$ [95,51]. HB: Harwell booster, ORL: Oak Ridge Electron LINAC, HSC: Harwell Sychrocyclotron, and KSFC: Karlsruhe Cyclotron. The numbers in brackets are the accelerator pulse length in ns.

At neutron energies above about 100 keV, organic scintillators[98] meet essentially all of the above requirements. The invention of pulse–shape discrimination by Brooks[98] makes it possible to reduce greatly the response of the system to photons in certain organic scintillators (notably stilbene), whilst leaving the response to neutrons almost unchanged. The method is most effective at neutron energies above about 500 keV, and at total counting rates of less than 50 kHz. Developments in the electronic methods used to separate the photon from the neutron pulses have been such that excellent systems are now available commercially.

At neutron energies below about 100 keV, the two most widely used detectors are the ^{10}B–NaI(Tl) combination, first developed by Thornton[99], and the ^{6}Li-loaded glass scintillator[100,101], developed from the original work of Ginther and Schulman[102] on cerium-activated glass scintillators. In the first of these detectors, the 480 keV γ-ray from the reaction ^{10}B(n, α, γ)^{7}Li + 2.3 MeV, is observed with a NaI(Tl) scintillation spectrometer. The efficiency of this system at low energies is limited largely by the solid angle covered by the NaI(Tl) crystal(s). The thermal neutron cross section for the primary reaction is almost 4000 b. In the glass scintillator, the reaction is ^{6}Li(n, α)T + 4.8 MeV, with a thermal neutron cross section of about 1000 barns. A glass scintillator with a thickness of 1″ has an efficiency of 17% at 1 keV. Both systems have intrinsic efficiencies that have a $1/v$-dependence on the neutron speed, v, in the keV-region.

The overall time resolution of a ^{10}B–NaI(Tl) detector can be made less than 3 ns, and in the ^{6}Li-loaded glass scintillator, an overall resolution of less than 5 ns has been achieved[103].

In addition to the detection of neutrons, the full range of photon and particle detectors has been, and is, used. These include NaI(Tl) and Ge(Li) spectrometers[104] for photon detection, solid state spectrometers for measurements of fission fragments[105], protons, deuterons and alpha particles, and gas scintillators[106], particularly for the detection of fission fragments.

6. Energy calibration of neutron time-of-flight spectrometers

6.1. DIRECT METHODS

The calibration of the time-scale of a time-of-flight spectrometer can be carried out directly against the period of a stable, crystal–controlled quartz oscillator with a typical frequency in the MHz range. Using fast scalers, synchronized pulse trains, with spacings ranging from tens of nanoseconds to seconds can be obtained readily. These pulse trains are used as the "stop" signals of the system, so that the channel-number scale can be converted directly into a time scale, related to the period of the master oscillator. If necessary, vernier techniques can be used to interpolate at the sub-nanosecond level, between the master clock pulses.

The fundamental period of the master oscillator can be calibrated in several ways: one method is to compare a sub-multiple of its frequency with a "standard" frequency from a radio transmitter broadcasting, for example, at a wavelength in the 1500 m range.

The production of a pulse of neutrons from an accelerator is invariably accompanied by a pulse of photons. The time-of-flight of the photons (traveling at 0.29979250 m/ns) can be recorded in the spectrometer and used as precise calibration point on the time scale.

For flight paths ranging in length from about 10 m to several hundred meters, standard surveying methods are used to determine the length. An accuracy of a few millimeters at the longest distances is achieved; these methods present no basic limitations at the present time.

It must be remembered that the rest energy of the neutron is currently not known to an accuracy of better than ± 1 keV.

6.2. INDIRECT METHODS

The problems associated with a direct calibration of the energy scale can be avoided by observing the time-of-flight spectrum of neutrons from a reaction that exhibits several clearly defined resonances at accurately known energies. These energies may have been obtained, initially, from a directly calibrated time-of-flight spectrometer, or from studies of (p, n) reactions in which the proton energy is determined with a magnetic spectrometer, calibrated with an NMR system.

In the MeV-region, resonances in the ^{12}C + n compound system are frequently used as energy standards (see table 7).

At lower energies, many well-known resonances can be used as standards; particularly useful are those listed by Wynchank[108] and used over the years in the extensive neutron time-of-flight stu-

TABLE 7

Resonances in the $^{12}C+n$ compound that are used as every standards in the MeV-region.

Resonance energy [107] (MeV)	Inverse speed (ns/m)
2.077 ± 0.002	50.248
2.950 ± 0.002	42.192
4.260 ± 0.020	35.147
6.293 ± 0.005	28.965
6.361 ± 0.005	28.811

dies at the Nevis cyclotron of Columbia University.

In the indirect method, it is, of course, not necessary to know the length of the flight path and the duration of the individual timing channels, independently. For example, if the timing system is linear (hopefully the case!), the inverse speed of a neutron is:

$$1/v_n = t_n/l = (\Delta t_n/l) N + (t_0/l),$$

where each channel has a duration Δt_n, $N = 0, 1, 2...$ labels the channel number, and t_0 is the "time-zero" delay. A least-squares fit to this linear relationship therefore gives the optimum values of $(\Delta t_n/l)$ and (t_0/l).

7. Summary

Remarkable developments in the field of neutron time-of-flight spectroscopy took place between 1950 and 1970. During that period, many different types of charged-particle accelerators were used to provide intense pulses of neutrons via appropriate reactions. Since that time, the method has continued to flourish but no significant breakthroughs have been reported... pulsed cyclotrons, linear electron and proton accelerators, Van de Graaff accelerators, Cockcroft–Walton generators and pulsed reactors had already reached a high level of performance. The pulse durations available were often in the range 1–10 ns (even as low as 50 ps), pulsed currents of electrons in excess of 10 A had been produced, and pulsed proton and deuteron currents of several A at repetition rates of several megahertz were available. Two main factors have slowed developments in the field during the last decade, they are: a general reduction in the amount of money available for the design and construction of particle accelerators for nuclear physics, and the natural difficulties of making major developments in an area that

is already highly developed. It is not a straightforward matter to go from a 2 ns, 10 A pulse of 100 MeV electrons or protons to a 1 ns, 100 A pulse of 1 GeV electrons or protons! ... and yet such an improvement in accelerator performance is necessary if we are to increase our understanding of the nature of neutron interactions with nuclei, particularly in the energy range from a few MeV to a few GeV. Improvements in both resolution and flux are essential if we are to carry out studies of those important and intriguing reactions with very small cross sections; such studies will be concerned not only with spectroscopy but also with questions of a fundamental nature such as the search for parity-violating parts of the interaction using polarized beams.

A modest extrapolation of our present knowledge of particle accelerators would indicate that, given reasonable encouragement, neutron time-of-flight spectrometers with resolutions of less than 1 ps per meter could be developed in the near future. During the course of such developments (particularly in the area of proton accelerators where the efficiency for generating neutrons by the spallation process is reasonable), we must not forget the need to develop particle detectors with even greater efficiencies and faster time responses than currently available.

Research in the field of laser-induced fusion may well lead to a totally new method for producing very intense bursts of neutrons with intrinsic durations of a few ps [109]. Intensities during the pulse could reach 10^{28} neutrons per second – a factor of 10^9 greater than those of present-day accelerator-based sources. Although the neutron energy would be 14 MeV, it has been estimated that there would be sufficient self-moderation in the highly compressed material to give a large flux of neutrons from the keV-region upwards [110].

References

[1] J. R Dunning, G. B. Pegram, G. A. Fink, D. P. Mitchell and E. Segré, Phys. Rev. **48** (1935) 704.

[2] E. Fermi, J. Marshall and L. Marshall, Phys. Rev. **72** (1947) 193.

[3] L. W. Alvarez, Phys. Rev. **54** (1938) 609.

[4] G. E. F. Fertel, D. F. Gibbs, P. B. Moon, G. R. Thompson and C. E. Wynn-Williams, Proc. Roy. Soc. **175** (1940) 316.

[5] C. P. Baker and R. F. Bacher, Phys. Rev. **59** (1941) 332.

[6] L. Cranberg and J. S. Levin, Phys. Rev. **103** (1956) 343.

[7] W. M. Good, J. H. Neiler and J. H. Gibbons, Phys. Rev. **109** (1958) 926.

562   F. W. K. FIRK

8) G. A. Kolstad, Ph. D. Thesis (Yale University, 1948).

9) J. D. Cockcroft, Nature 163 (1949) 869.

10) M. L. Yeater, E. R. Gaerttner and G. C. Baldwin, Rev. Sci. Instr. 28 (1957) 514.

11) B. N. Taylor, W. H. Parker and D. N. Langenberg, Rev. Mod. Phys. 41 (1969) 375.

12) H. A. Bethe, Rev. Mod. Phys. 9 (1937) 69.

13) S. Cierjacks, B. Duelli, P. Forti, D. Kopsch, L. Kropp, M. Losel, J. Nebe, H. Schweikert and H. Unseld, Rev. Sci. Instr. 39 (1968) 1279.

14) J. B. Garg, L. J. Rainwater and W. W. Havens Jr., Phys. Rev. 137B (1965) 547.

15) G. Breit and E. P. Wigner, Phys. Rev. 49 (1936) 519 and 642.

16) H. A. Bethe and G. Placzek, Phys. Rev. 51 (1937) 450.

17) S. E. Atta and J. A. Harvey, J. Soc. Indust. Appl. Math. 10 (1962) 617.

18) C. E. Porter and R. G. Thomas, Phys. Rev. 104 (1956) 483.

19) E. P. Wigner, Proc. Gatlinburg Conf. on Neutron time-of-flight methods, Oak Ridge Natl. Lab. Rept. ORNL. 2309 (1957).

20) M. L. Mehta, Nucl. Phys. 18 (1960) 395.

21) R. O. Lane, Nuclear structure studies with neutrons (eds. J. Erö and J. Szücs; Plenum Press, London, 1974) p. 31.

22) S. Cierjacks, P. Forti, D. Kopsch, L. Kropp, J. Nebe and H. Useld, Karlsruhe, W. Germany, Rept. KFK 1000 (1968).

23) C. K. Bockelman, D. W. Miller, R. K. Adair and H. H. Barschall, Phys. Rev. 84 (1951) 69.

24) C. M. Turner and S. D. Bloom, Rev. Sci. Instr. 29 (1958) 480.

25) J. H. Neiler and W. M. Good, Fast neutron physics, part 1 (eds. J. B. Marion and J. L. Fowler; Interscience Publ., New York, 1960) p. 509.

26) K. R. Spangenberg, Vacuum tubes (McGraw-Hill, New York, 1948).

27) V. J. Ashby, W. M. Harris, A. F. Klein and I. Nakada, Univ. of Calif. Rad. Lab. Rept. UCRL 4641 (1955).

28) N. N. Flerov and E. A. Tamanov, Atom. Energ. 3 (1957) 44; J. Nucl. En. 8 (1958) 91.

29) C. Delaney, At. En. Res. Est. Harwell, England, Rept. NP/R 1666 (1955).

30) ORTEC Inc., Oak Ridge, TN. U.S.A.

31) R. C. Mobley, Phys. Rev. 88 (1952) 360.

32) A. T. G. Ferguson, N. Gale, G. C. Morrison and R. E. White, Direct interactions and nuclear reaction mechanisms (Gordon and Breach, New York, 1963) p. 510.

33) D. Bohm and L. Foldy, Phys. Rev. 72 (1947) 649.

34) H. J. S. Allwood, Ph. D. Thesis (Birmingham University, 1948).

35) B. L. Cohen, Rev. Sci. Instr. 24 (1953) 589.

36) G. A. Fink, Phys. Rev. 50 (1936) 738.

37) T. Brill and H. V. Lichtenberger, Phys. Rev. 72 (1947) 585.

38) W. Selove, Phys. Rev. 84 (1951) 869.

39) P. A. Egelstaff, Neutron time-of-flight methods (ed. J. Spaepen; European Atomic Energy Community (Euratom), Brussels, 1961) p. 261.

40) R. C. Block, G. G. Slaughter and J. A. Harvey, Nucl. Sci. Eng. 8 (1960) 112; R. T. Carpenter and L. M. Bollinger, Nucl. Phys. 21 (1960) 66.

41) L. J. Rainwater and W. W. Havens, Jr., Phys. Rev. 70 (1946) 136.

42) W. W. Havens, Jr., and L. J. Rainwater, Phys. Rev. 70 (1946) 154.

43) L. J. Rainwater, W. W. Havens, Jr., and J. B. Garg, Rev. Sci. Instr. 35 (1964) 263.

44) E. Hacken, L. J. Rainwater, H. I. Liou and U. N. Singh, Phys. Rev. 13 (1976) 1884.

45) P. H. Bowen, J. P. Scanlon, G. M. Stafford and J. J.Thresher, Nucl. Phys. 22 (1961) 640.

46) A. Langsford, P. H. Bowen, G. C. Cox, F. W. K. Firk, D. B. McConnell and B. Rose, Nuclear structure study with neutrons (eds. M. Neve de Mevergnies, P. Van Assche and J. Vervier; North-Holland Publ. Co., Amsterdam, 1966) p. 562.

47) A. E. Taylor, private communication.

48) R. G. P. Voss and R. Wilson, Phil. Mag. 1 (1956) 175.

49) P. Hillman, G. H. Stafford and C. Whitehead, Nuovo Cim. 4 (1956) 67.

50) S. Cierjacks, Nuclear structure studies with neutrons (eds. J. Erö and J. Szücs; Plenum Press, London, 1974) p. 299.

51) C. C. Cutler, Bell Tel. Lab. Rept. MM/44/160/218 (1944).

52) D. W. Fry, R. S. Harvie, L. Mullett and W. Walkinshaw, Nature 160 (1947) 351.

53) E. L. Ginzton, W. W. Hansen and W. R. Kennedy, Rev. Sci. Instr. 19 (1948) 89.

54) H. L. Schultz, E. R. Beringer, C. C. Clarke, J. A. Lockwood, R. L. McCarthy, C. G. Montgomery, P. J. Rice and W. W. Watson, Phys. Rev. 72 (1947) 346A; H. L. Schulz and W. G. Wadey, Rev. Sci. Instr. 22 (1951) 383.

55) J. C. Slater, Phys. Rev. 70 (1946) 799A and Rev. Mod. Phys. 20 (1948) 473.

56) C. A. Fenstermacher, R. G. Bennett, A. E. Walters, C. K. Bockelman and H. L. Schultz, Phys. Rev. (1957) 1650.

57) W. Bertozzi, C. P. Sargent and W. Turchinetz, Phys. Lett. 6 (1965) 108.

58) C. F. Bareford and M. G. Kelliher, Phillips Tech. Rev. 15 (1953) 1.

59) E. R. Rae, E. R. Collins, B. B. Kinsey, J. E. Lynn and E. R. Wiblin, Nucl. Phys. 5 (1958) 89.

60) H. H. Landon and E. R. Rae, Phys. Rev. 107 (1957) 1333.

61) F. W. K. Firk, Nucl. Phys. 9 (1958) 198.

62) M. J. Poole, J. Nucl. En. 5 (1957) 325.

63) F. W. K. Firk, G. W. Reid and J. F. Gallagher, Nucl. Instr. 3 (1958) 309.

64) F. W. K. Firk, J. E. Lynn and M. C. Moxon, Nucl. Phys. 41 (1963) 614.

65) M. J. Poole and E. R. Wiblin, Proc. 2nd. Int. Conf. on Peaceful uses of atomic energy, Geneva, vol. 14 (1958) p. 266.

66) F. W. K. Firk and E. M. Bowey, Comptes Rendus du Congres Intern. de Physique nucléaire (ed. P. Gugenberger; C.N.R.S., Paris, 1964) p. 1013.

67) R. Bergère, Neutron time-of-flight methods (ed. J. Spaepen; European Atomic Energy Community (Euratom), Brussels, 1961) p. 329.

68) A. Michaudon and P. Ribon, ibid., p. 565.

69) O. A. Wasson and J. E. Draper, Nucl. Phys. 73 (1965) 499.

70) F. W. K. Firk, Nucl. Instr. and Meth. 43 (1966) 43.

71) E. R. Gaerttner, M. L. Yeater and R. R. Fullwood, Neutron physics (Academic Press, New York, 1962) p. 263.

72) D. R. Adcock, J. R. Beyster and J. L. Cole, IEEE Trans. Nucl. Sci. NS-14 (1967) 721.

73) J. A. Harvey, Proc. Int. Conf. on Interactions of neutrons with nuclei (ed. E. Sheldon; USERDA Tech. Info. Center, Oak Ridge, TN., 1976) p. 144.

74) F. L. Shapiro, Nuclear structure study with neutrons (eds. M.

Neve de Mevergnies, P. Van Assche and J. Vervier; North-Holland Publ. Co., Amsterdam, 1966).

[75] G. A. Keyworth, J. R. Lemley, C. E. Olsen, F. T. Seibel, J. W. T. Dabbs and N. W. Hill, Phys. Rev. C8 (1973) 2352.

[76] N. J. Norris and R. K. Hanst, Edgerton, Germeshausen and Grier, Prog. Rept. EGG 1183-2142 (1967).

[77] R. M. Laszewski, R. J. Holt and H. E. Jackson, Phys. Rev. Lett. 38 (1977) 813.

[78] S. D. Bloom, Nuclear reactions, vol. 2 (eds. P. M. Endt and P. B. Smith; North-Holland Publ. Co., Amsterdam, 1962) p. 1.

[79] J. D. Anderson and C. Wong, Phys. Rev. Lett. 7 (1961) 250.

[80] L. Cranberg and J. S. Levin, Phys. Rev. 100 (1955) 434.

[81] R. Batchelor and J. H. Towle, Proc. Phys. Soc. 73 (1959) 193.

[82] W. M. Good, Neutron time-of-flight methods (ed. J. Spaepen; European Atomic Energy Community (Euratom), Brussels, 1961) p. 309.

[83] J. H. Gibbons, R. L. Macklin, P. D. Miller and J. H. Neiler, Phys. Rev. 122 (1961) 182.

[84] F. W. K. Firk and J. H. Gibbons, Neutron time-of-flight methods (ed. Spaepen; European Atomic Energy Community (Euratom), Brussels, 1961) p. 213.

[85] R. L. Macklin and J. H. Gibbons, Rev. Mod. Phys. 37 (1965) 166.

[86] Nuclear research with low energy accelerators (eds. J. B. Marion and D. M. Van Patter; Academic Press, New York, 1967) pp. 141, 167 and 359.

[87] Y. S. Yazvitskii, Nuclear structure study with neutrons (eds. J. Erö and J. Szücs; Plenum Press, London, 1974) p. 335.

[88] F. L. Shapiro, ibid., p. 259.

[89] M. S. Moore, ibid., p. 327.

[90] E. R. Rae, Fundamentals of nuclear theory, (IAEA, Vienna, 1967), p. 831.

[91] A. Langsford, private communication.

[92] F. W. K. Firk, Proc. Int. Nuclear Physics Conf. (ed. R. L. Becker; Academic Press, New York, 1967) p. 352.

[93] E. R. Collins, private communication.

[94] P. Ribon and A. Michaudon, Neutron time-of-flight methods (ed. J. Spaepen; European Atomic Energy Community (Euratom), Brussels, 1961) p. 357.

[95] E. R. Rae and W. M. Good, Experimental neutron resonance spectroscopy (ed. J. A. Harvey; Academic Press, New York, 1970) p. 1.

[96] J. E. Lynn, private communication.

[97] G. A. Bartholomew and P. R. Tunicliffe, Chalk River Rept. AECL-2600 (1966).

[98] F. D. Brooks, Progress in nuclear physics, vol. 5 (ed. O. R. Frisch; Pergamon Press, Oxford, 1956) p. 252.

[99] W. A. Thornton, Ph. D. Thesis (Yale University, 1951).

[100] V. K. Voitovetskii, N. S. Tolmacheva and M. N. Arseav, Atom. Energ. 6 (1959) 321.

[101] F. W. K. Firk, G. G. Slaughter and R. J. Ginther, Nucl. Instr. and Meth. 13 (1961) 313.

[102] R. J. Ginther and J. H. Schulman, IEEE Trans. Nucl. Sci. NS-15 (1958) 92.

[103] F. D. Brooks, Neutron time-of-flight methods (ed. J. Spaepen; European Atomic Energy Community (Euratom), Brussels, 1961) p. 389.

[104] K. J. Wetzel, Ph. D. Thesis (Yale University, 1965).

[105] E. Melkonian, Nucl. Instr. and Meth. 11 (1961) 307.

[106] W. W. Havens, Jr., E. Melkonian, L. J. Rainwater and J. L. Rosen, Phys. Rev. 116 (1959) 1538.

[107] S. F. Mughabghab and D. I. Garber, Neutron cross sections, vol. 1, BNL 325 3rd edition. (Natl. Tech. Info. Service, Springfield, VA, 1973).

[108] S. A. R. Wynchank, Proc. Conf. on Neutron cross section technology (ed. P. B. Hemmig; USAEC Tech. Info. Center, Oak Ridge, TN., 1966) p. 287.

[109] E. K. Storm, H. G. Ahlstrom, M. J. Boyle, L. W. Coleman, H. N. Kornblum, R. A. Lerche, D. R. MacQuigg, D. W. Phillion, F. Rainer, V. C. Rupert, V. W. Slivinsky, D. R. Speck, C. D. Swift and K. G. Tirsell, Phys. Rev. Lett. 40 (1978) 1570.

[110] J. E. Lynn, Proc. Int. Conf. on Interactions of neutrons with nuclei (ed. E. Sheldon; USERDA Tech. Info. Center, Oak Ridge, TN., 1976) p. 287.

N-P CAPTURE AND THE PHOTO-DISINTEGRATION OF THE DEUTERON

F. W. K. Firk

Yale University

New Haven, Ct. 06520, U. S. A.

Abstract: Recent experiments on n-p radiative capture at thermal energies, including the study of double-photon decay, and on the photo-disintegration of the deuteron are discussed. Comparisons between the results and contemporary theories are made whenever they are meaningful. Suggestions for some necessary experiments for the future are made.

INTRODUCTION

From the very beginning of Nuclear Physics, studies of the radiative capture of thermal neutrons by protons and, particularly, of the photo-disintegration of the deuteron have played a significant role quite beyond anything that the pioneers in the field could have foreseen. We shall see that remarkable developments continue to take place in the field, almost annually.

In 1934, Chadwick and Goldhaber[1] not only observed the photo-disintegration of the deuteron (the "diplon") using photons from a thorium-C source (photon energy about 2.6 MeV) to irradiate an ionization chamber containing deuterium, but also improved the measurement of the mass of the neutron by estimating the energy of the photoprotons from their ionization. They used the principle of detailed balance to obtain the cross section for the capture of neutrons by protons, and concluded that it is much smaller than the photo-disintegration cross section. Almost simultaneously, Bethe and Peierls[2] published an account of the interaction of radiation with the two-nucleon system that included a discussion of photo-disintegration, of n-p capture and, for good measure, they considered the electro-disintegration of the deuteron. As early as 1935, Fermi[3] showed that in n-p capture at very low energies, capture from S-states in the continuum is important ($^1S \rightarrow ^3S$), resulting in a dominant M1 transition. Further theoretical developments

took place in the next decade, for example, Breit and Condon[4] introduced a nucleon-nucleon force that included a radial dependence and
both exchange and non-exchange parts. Rarita and Schwinger[5] introduced non-central forces, and found that they had a considerable
effect on the angular distribution of the photonuclons. DeSwart
et al.[6] and Zernik et al.[7] used hard-corenucleon-nucleon po-
tentials with spin-orbit and tensor components, and thereby accounted for the large isotropic part of the observed angular distribution of photoprotons. Refinements were introduced, particulary
by Partovi[8] and by Nunemaker,[9,10] that pushed the classical non-
relativistic Schrödinger-type calculations to their limit. These
calculation were taken very seriously during the 1960's (too
seriously by many workers in the field!) and, even today, are widely
used for comparisons with experimental data. In recent years,
theories have been developed that go far beyond the Partovi-
Nunemaker variety; Le Bellac et al.[11] used dispersion-theoretic
techniques to obtain cross sections in reasonable agreement with
experiment, and explicit inclusion of meson exchange currents
were made by Adler et al.,[12] Gari and Huffman,[13] and Riska and
Brown.[14] The latter workers found that the longstanding discrepancy
between the measured value of 334.2 ± 0.5 mb obtained by Cox et al.,[15]
and the "best" (non-meson) theoretical value of 302 ± 4mb[16] for the
n-p capture cross section of thermal neutrons, could be accounted
for by inclusion of pion exchange operators (see Villars[17]) and
those involving the N_{33} isobar and the process $\omega \rightarrow \pi\gamma$. They also
demonstrated the critical role played by the inclusion of matrix
elements between the ^1S continuum state and the $^3D_1$ component of
the deuteron; this fact seems to have been overlooked in previous
calculations.

Hadjimichael[18] made a direct calculation of the effects of
meson exchange processes in the photo-disintegration of the deuteron
at energies below about 20 MeV. He concluded that such processes
have a very small effect on the total cross section, and on the
differential cross section but they can have measurable effects on
the polarization of the photonucleons. This point will be taken
up later.

Arenhövel, Fabian and Miller,[19] have shown the importance not
only of meson exchange currents but also of the NΔ, NN and ΔΔ isobar
configurations on the ground state wave function of the deuteron.

Renewed interest in the standard calculation of the dominant
E1-component of the interaction at photon energies above about 5 MeV
has come about following the recent work of Gari and Sommer[20] in
generalizing the electric dipole operator to include, correctly,
so called exchange terms, that were not included in the famous work
of Levinger and Bethe.[21] Their conclusions will be presented in
section III. Yet another major theoretical development has appeared
within the last eighteen months to explain the remarkable resonance
observed in the polarization of photoprotons at an photon energy of
530 MeV;[22] this resonance is attributed to a new state of matter -
a ΔΔ dibaryon state.[23] This work is discussed in section IV.

I.   THE THERMAL NEUTRON ABSORPTION CROSS SECTION OF HYDROGEN
     Three methods have been used throughout the years to measure
n-p capture cross section at thermal energies, they are
  i) the steady state method in which the spatial distribution
     of neutrons is measured in a large hydrogenous material,
     and then in an homogeneous standard absorber (eg: boron).
     The method is limited by the accuracy of the value of the
     boron cross section, and the ^{10}B content.
 ii) the oscillatory method in which the response of a critical
     reactor to a periodically varying absorber is measured.
     The cross sections of standard absorbers must again be
     known.
iii) the pulsed source method in which the capture cross section
     is determined from the rate of exponential decay of the
     thermalized neutron population following short bursts of
     fast neutrons into water samples of various dimensions.
     The results of 21 independent measurements of the thermal n-p
capture cross section carried out using the different methods outlined above, and spanning a period of more than 30 years, are shown
in Figure 1. (References to the data are given in ref. 24). The
least-squares weighted value is 332.66 ± 0.36 mb. The distribution
of the values is plotted in Figure 2; two distinct groups are observed centered at 330 and 335 mb. The most recent measurement,
obtained by Cokinos and Melkonian[24] using the high-intensity pulsed
neutron source associated with the old 380-MeV Nevis cyclotron with
a wide variety of sample sizes, is 332.6 ± 0.7 mb. It is desirable
to make additional high precision measurements to corroborate the
mean value.
     As mentioned in the introduction, early attemps to calculate
the thermal n-p capture cross section using non-meson theories, gave
maximum values of 302 ± 4 mb. It was not until the work of Riska
and Brown[14] and, independently, of Gari and Huffman,[13] that the
essential terms involving pion exchange and excitation of isobars
were correctly included, and accounted for the bulk of the missing
cross section.

     II.   DOUBLE-PHOTON DECAY IN n-p CAPTURE AT THERMAL ENERGIES
     The earlier discrepancy between the observed n-p capture cross
section at thermal energy, and the predicted value obtained using
non-meson theory, prompted Arnold et al.[25] to study the reaction

$$n + p \rightarrow \gamma_1 + \gamma_2 + d$$

in order to see whether such a mechanism could explain the discrepancy. They searched for coincident photons with a total energy
of 2.22 MeV, emerging from a sample of H_2O irradiated with thermal
neutrons from the NBS reactor. They concluded that the ratio of
the cross section for producing two photons ($\sigma_{2\gamma}$) to that for producing one photon ($\sigma_{1\gamma}$) is:

$$\sigma_{2\gamma} / \sigma_{1\gamma} \stackrel{\sim}{<} 3 \times 10^{-3}$$

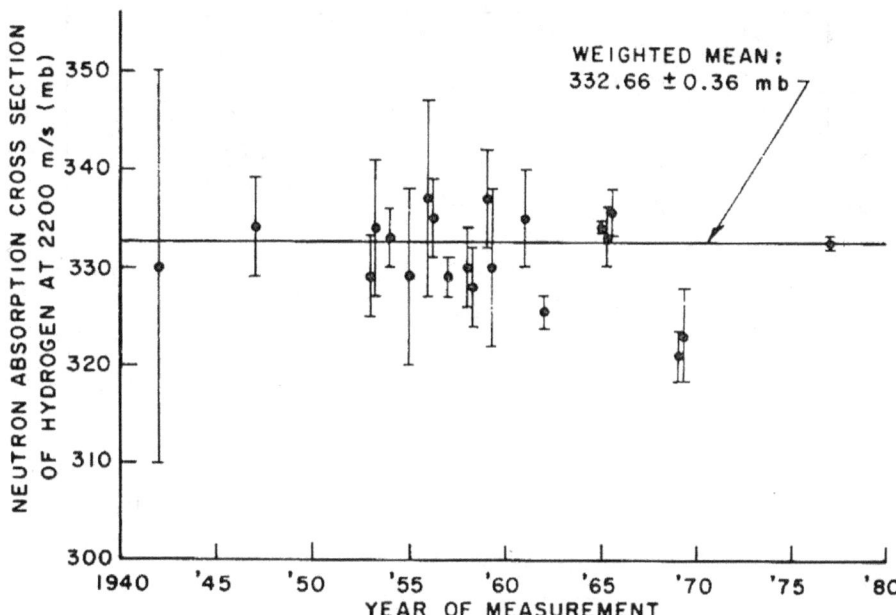

248   F. W. K. FIRK

Fig. 1.   The world values of the n-p capture cross section
for thermal neutrons. The data, with all references, are
given by Cokinos and Melkonian.[24]

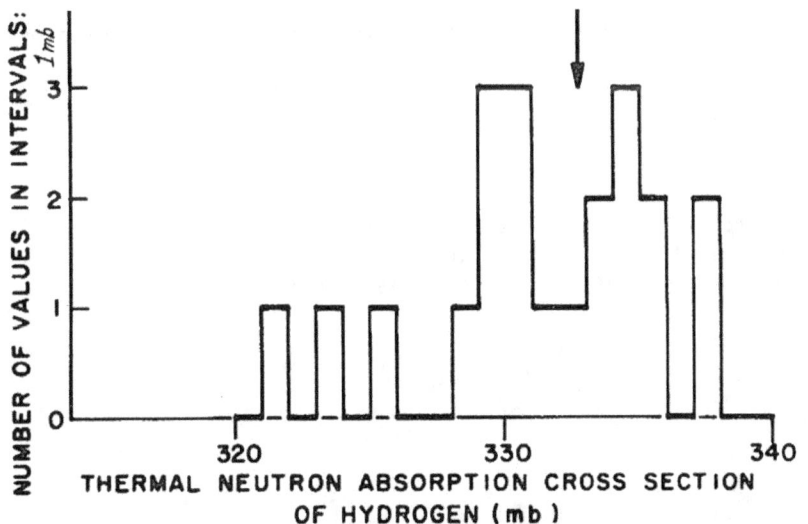

Fig. 2.   The distribution of world values of the n-p capture
cross section for thermal neutrons.

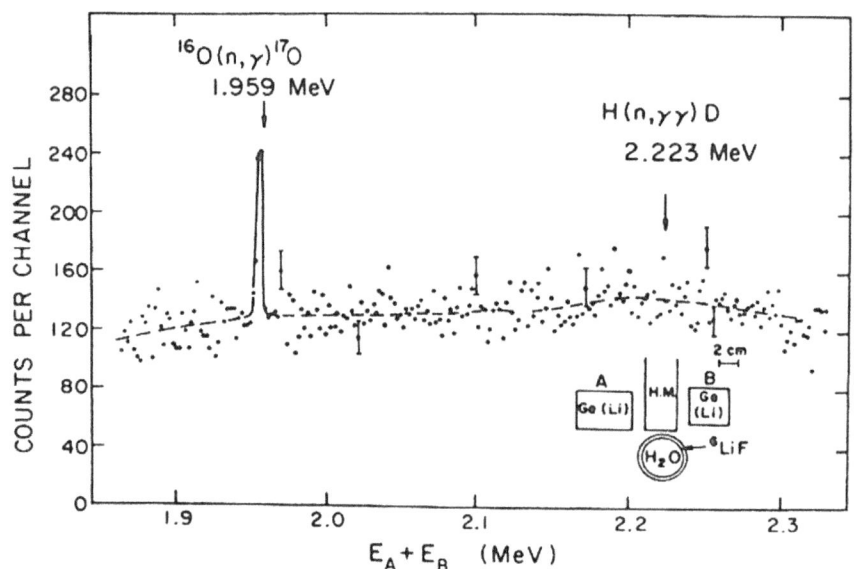

Fig. 3.   The observed spectrum of coincident photons with energies
$E_A + E_B$, using the geometry shown. The sum of the cascade
in ^{16}O + n, corresponding to a total energy of 1.959 MeV,
is used as a calibration of the system.

They noted that, in an earlier calculation, Adler[26] had estimated,
theoretically:

$$\sigma_{2\gamma} / \sigma_{1\gamma} = 3.4 \times 10^{-10} \qquad \text{(from the } ^1S_0 \text{ np capture state)}$$

and

$$\sigma_{2\gamma} / \sigma_{1\gamma} = 4.8 \times 10^{-13} \qquad \text{(from the } ^3S_1 \text{ state)}$$

Arnold et al.[25] concluded that the double-photon decay process
did not explain the discrepancy (how right they were!).
    In a later experiment, Dress et al.[26] reported the value

$$\sigma_{2\gamma} / \sigma_{1\gamma} = (1.05 \pm 0.16) \times 10^{-3}$$

which gives $\sigma_{2\gamma}$ = 350 ± 50 μb, using $\sigma_{1\gamma}$ = 334mb.
    Subsequent experiments[27,28] have shown the value obtained by
Dress et al.[26] to be in error (due to poor experimental geometry)
and that, using the system shown in Figure 3, Earle et al.[28] obtained a value of

$$\sigma_{2\gamma} / \sigma_{1\gamma} \leq 10^{-4}$$

    In a contribution to this conference, Earle and McDonald[29]
have set an upper limit of 1.6 μb for the p(n,γγ)d cross section for

F. W. K. FIRK

photons above 700 keV. They used two NaI detectors with a coincidence resolving time of 2.6 ns. The experiment took six months
(neutron flux 2×10^6 neutrons $cm^{-2} s^{-1}$), events were stored on
magnetic tape, event-by-event, and the data were analyzed in the
region of real coincidence events, stored in a 64 x 64 channel
matrix. They conclude that, in order to obtain a limit significantly
closer to the theoretical prediction of $\sigma_{2\gamma} \sim 0.07$ µb, major technical
advances in γ-ray detectors and/or their time resolution are needed.

III. THE PHOTO-DISINTEGRATION OF THE DEUTERON
       A. Differential Cross Sections.
       At the present time, no absolute measurements of the differ-
ential cross sections for the reaction d(γ,n)p exist. The problems
associated with studying fast neutrons with high resolution and with
precisely known efficiency are so formidable that only relative
measurements have been made. However, many studies of the d(γ,p)n
reaction have been made over a wide range of angles, and the
findings continue to generate controversy among experimenters and
gnashing of teeth among theorists. In my opinion, at photon energies
above about 15 MeV, only two experiments will genuinely stand the
test of time - others may have obtained the correct result but they
will all have relied on an element of good luck. The two experiments
that I refer to are i) the measurement of the differential cross
sections for the d(γ,p)n reaction between 25 and 55 MeV made by
Weissman and Schultz[30] and ii) the measurement of the 0° differential cross section between 30 and 120 MeV made by Hughes et al.[31]
       The Weissman-Schultz experiment represents all that is best
in our art; no stone was left unturned. The experiment went on for
a period of more than three years - a year to construct and test the
thick lithium-drifted silicon proton counters and to build the target
chamber, and two years in which to collect the data. This part of
their work was of the utmost importance - they carried out six
independent experiments widely separated in time. Sources of
systematic errors were investigated over this long period; often,
completely different experimental configurations were used. The
maximum energy was not the same from one run to another, different
photon beam-hardeners and photon collimators were used and, indeed,
completely different electronics were tested and set-up for each
measurement (typically, one complete measurement took 2 to 3
weeks of dedicated electron LINAC time). The data were analyzed
and studied carefully between runs. In this way, their final
values do not suffer from the traditional problems of systematic
errors that can plague the one-off experiment, even when great
care has apparently been taken. My only reservation concerning
the Weissman-Schultz work centers around their "25-MeV point"
which they themselves were dubious about; it is at the extreme
lower-end of the energy range where an adequate signal-to-noise
ratio could be obtained. Their results are shown in Figure 4,
together with some earlier results of Allen.[32] The dashed curves

180

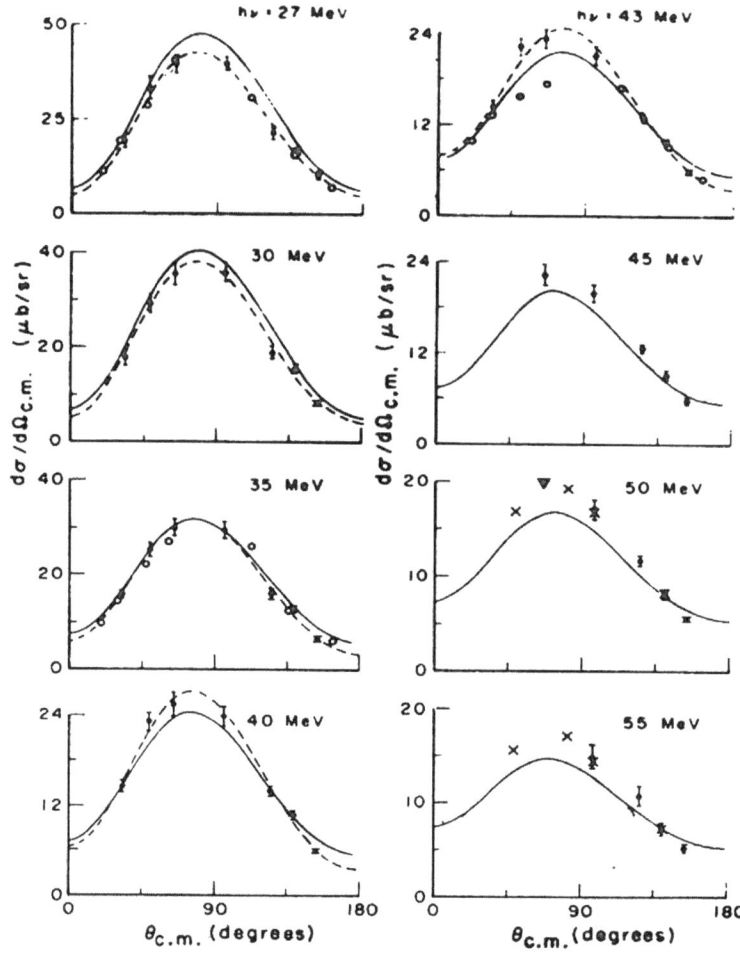

Fig. 4. The values of the differential cross sections for the
reaction d(γ,p)n obtained by Weissman and Schultz[30]
(closed circles) and by Allen[32] (open circles). The
dashed lines are fits to the data using a Legendre expansion, and the solid curves are the predictions of
Partovi.[8]

TABLE 1

THE ANGULAR DISTRIBUTION COEFFICIENTS FOR THE REACTION

$$\gamma + d \rightarrow n + p$$

IN THE PHOTON ENERGY RANGE 27 – 40 MEV. [30]

PHOTON ENERGY (MeV)	A_0 ($\mu b.sr^{-1}$)	A_1 ($\mu b.sr^{-1}$)	A_2 ($\mu b.sr^{-1}$)	A_3 ($\mu b.sr^{-1}$)	A_4*
27	29.1±0.8	6.33±0.89	-23.3±1.6	-7.4±2.0	100%
30	25.9±0.7	6.07±0.84	-21.2±1.5	-5.9±1.6	100%
35	21.8±0.6	5.53±0.57	-17.3±1.2	-4.5±1.3	60%
40	18.9±0.5	5.31±0.45	-13.9±1.0	-3.8±1.0	30%

* PROBABILITY OF NOT BEING NECESSARY

TABLE 2

THE RESULTANT DIFFERENTIAL CROSS SECTIONS AT

0° AND 180° OBTAINED FROM THE ABOVE COEFFICIENTS

PHOTON ENERGY (MeV)	$\sigma(\theta_{CM}=0^{\circ})$ $(\mu b.sr^{-1})$	$\sigma(\theta_{CM}=180^{\circ})$ $(\mu b.sr^{-1})$
27	4.7 ± 1.4	6.9 ± 2.0
30	4.9 ± 1.5	4.5 ± 1.4
35	5.5 ± 1.7	3.5 ± 1.0
40	6.6 ± 2.0	3.4 ± 1.0

represent analytical fits to the Weissman-Schultz data using the familiar form

$$\sigma(\theta) = \sum_{\ell=0}^{n} A_\ell P_\ell (\cos\theta)$$

The solid curve represents the predictions of Partovi.[8] At all energies, the data disagree with the calculation. Below 35 MeV, the calculated cross sections are too large, and above that energy, they are too small. The Legendre coefficients obtained by Weissman and Schultz[30] are listed in Table 1. It is of interest to calculate the 0° and 180° differential cross sections from their coefficients: they are listed in Table 2. At energies up to 35 MeV, the cross section at 0° is less than $5.5 \mu b.sr^{-1}$. At 40 MeV, the increased value of $6.6 \mu b.sr^{-1}$ can be attributed to the onset of a contribution corresponding to the A_4 coefficient, which has a 70% probability of being required at that energy. It was gratifying to find that the measurements of the 0°-differential cross section, reported by Hughes et al.[31] in 1976 (see Figure 8), demonstrated conclusively that the earlier values of Weissman and Schultz, in the region of 30 MeV, were indeed correct. Let us hope that future measurements at 180° will also confirm the values listed in Table 2. Comparisons between the measured values of the Legendre coefficients and the predictions of Partovi are presented in Figure 5. The general agreement is reasonable, but the detailed agreement is poor; for the past decade we have therefore been aware of the shortcomings of the Partovi calculation.

The measurements of Baglin et al.[33], made at the Electron Prototype Accelerator at Los Alamos in the late 60's, cover the range from 17 to 25 MeV. They therefore obtain only one point of overlap with the Weissman-Schultz data at 25 MeV, and as mentioned earlier, this single point is not totally reliable. The main problem with the work of Baglin et al.[33] stems from the fact that it was a precision experiment (± 6%) carried out once, and once only. The experiment was not repeated (the accelerator was closed down in the middle of their work) so that it was not possible to investigate possible systematic errors. With this major objection in mind, it is difficult to get too excited about the discrepancies between their measured Legendre coefficients and those predicted by Partovi. It is interesting to note that the values of the $0°$-differential cross section deduced from the Baglin et al. Legendre coefficients below 20 MeV are typically less than 3 µb.sr^{-1} - values that are at variance with the most recent calculations of Gari and Sommer,[20] see Figure 8. Precise measurements of the $0°$- and $180°$-differential cross sections at low energies are essential in placing constraints on the values of the (small number of) Legendre coefficients needed below about 20 MeV.

Measurements of the relative angular distributions of photo-neutrons from the reaction d(γ,n)p yield ratios of Legendre coefficients that are in very good agreement with those obtained by Weissman and Schultz. The measurements of Shin et al.[34] are shown in Figure 6. Again, appreciable differences are found between the measurements and the Partovi predictions over the entire photon range from about 20 to 100 MeV.

B.   Total Absorption Cross Section

Only one measurement of the total photon absorption cross section of the deuteron has been reported in the region 20 to 30 MeV, and that was made using a D_2O - H_2O difference method. The nuclear absorption in water represents <0.1% of the total absorption processes in water! Nonetheless, Ahrens et al.[35] state that they have measured the cross section for nuclear photo-disintegration of the deuteron with a systematic error of only 2.9% (at 20 MeV). This work suffers from the same basic problem that I find in the work of Baglin et al. namely, it is a one-off experiment. All manner of systematic errors can enter such a transmission experiment - how does the result depend on absorber thickness? - how are the backgrounds measured? - how does the result depend on the form of deuterium? - have gas targets of various pressures been used? - have pure liquid targets of various lengths been tried? . . etc.   It is difficult to assess the reliability of the result when we are presented with no more than two pages in a Physics Letter! Clearly, many more independent measurements are required before we can claim to have definitive values of the total cross section throughout the range 2-100 MeV (say).

The total cross sections obtained by Weissman and Schultz by integrating-over-angles are compared with the classical calculation of Partovi in Figure 5. At energies above 35 MeV, their values are greater than the predicted values, at 35 MeV it is in exact agreement with the calculated value, and at two energies below 35 MeV, the measured cross section is less than the Partovi value (their 25-MeV point should be treated with appropriate skepticism). The total cross sections deduced by Baglin et al. are about 20% smaller

184

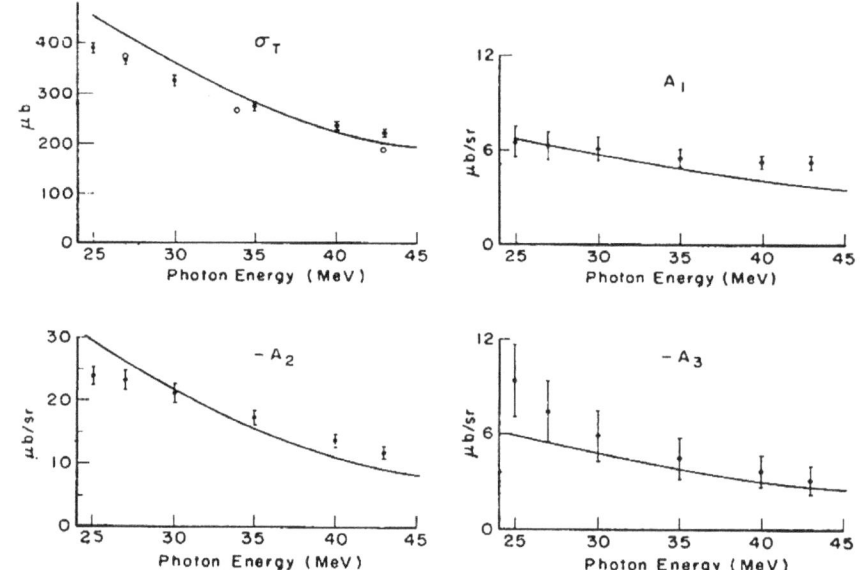

Fig. 5   Comparisons between the Legendre coefficients obtained
            by Weissman and Schultz[30], and the predictions of
            Partovi.[8]   Note that $\sigma_T = 4\pi A_o$.

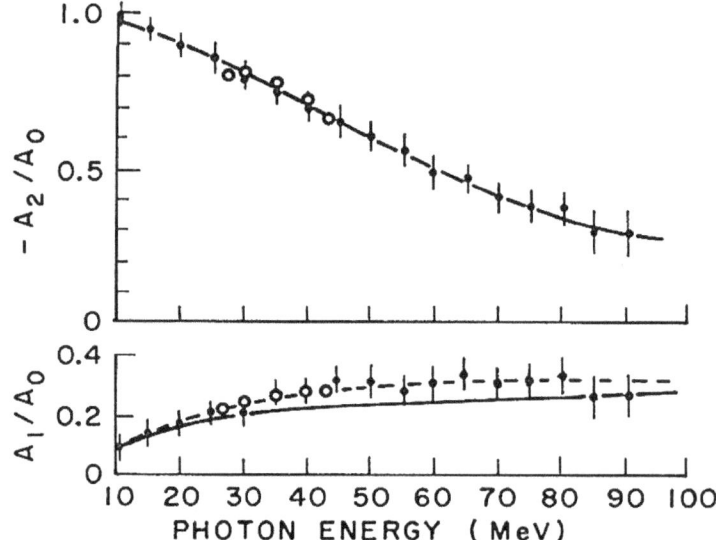

Fig. 6.   The coefficients A_1, A_2 compared with A_o, for
            the reaction $d(\gamma,n)p$ (Shin et al.[34]   The open circles
            are the values obtained in ref. 30, and the solid curves
            are those of Partovi.[8]

256   F. W. K. FIRK

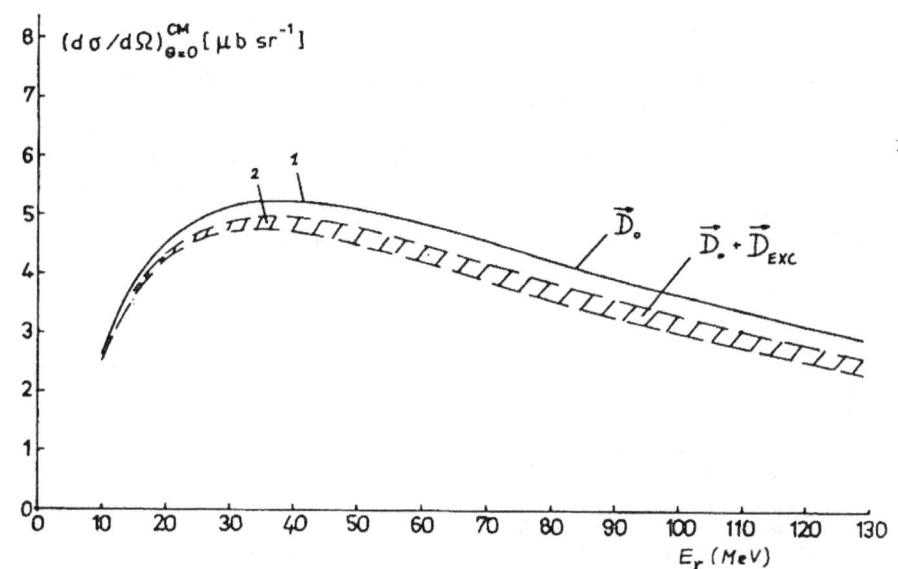

Fig. 7.   The El-component of the 0°-differential cross section
           for the reaction d(γ,p)n calculated by Gari and Sommer[20]

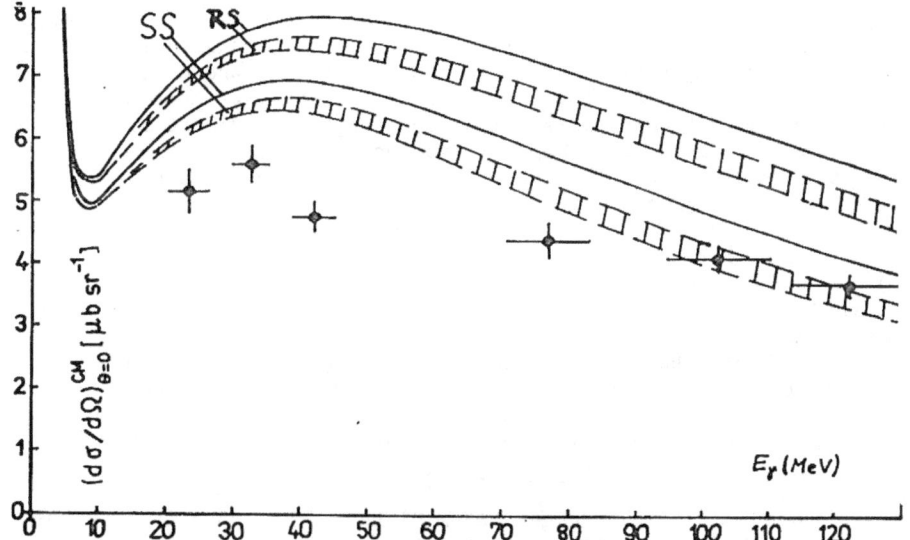

Fig. 8.   The complete calculation of the 0°-differential cross
           section for the reaction d(γ,p)n calculated by Gari and
           Sommer[20] using the general dipole operator for a Reid
           soft-core potential (RS) and a super-soft core potential
           (SS).

than "Partovi" at 20 MeV and less than 10% smaller at 17 MeV.
Baglin et al. concluded that their results could be interpreted
as an indication of an error in the basic El component of the theory!
Recent theoretical work by Gari and Sommer[20] indeed supports such
a notion - although it is unlikely to lead to such a dramatic 20%
effect on the total cross section at 20 MeV. Gari and Sommer consider not only the correction (κ) to the classical dipole sum rule
due to a non-vanishing spatial exchange current (Bethe-Levinger)
but also a non-vanishing two-body charge density (δ). They give:

$$\int_o \sigma_\gamma (w) \, dw \, \Big|_{El} = 60 \, (NZ/A)(1 + \kappa + \delta)$$

They conclude that present experiments and theories would
allow $|\delta| \sim 0.2$.

They have recalculated the deuteron photo-disintegration cross
section at 0° with dipole operators D_o (Siegert) and $D_o + D_{exc}$.
Their results are shown in Figures 7 and 8; the importance of this
work is clearly evident.

    C. Polarization Effects.

      1. Low energies

Studies of the polarization of photoneutrons from the reaction
$d(\gamma, \vec{n})p$ can provide information on the role of meson currents
even at energies well below the meson threshold. Hadjimichael[18]
and Arenhövel et al.[19] have shown that, whereas the energy dependence of the total and differential cross section for the
$d(\gamma,n)p$ reaction remain almost unchanged by the inclusion of meson
exchange currents and nucleon isobars, the differential polarization
can be changed appreciably. In both calculations, the effect of
meson currents and of isobars is to increase the magnitude of the
neutron polarization compared with those values predicted by the
classical calculations of Partovi and Nunemaker.

The predominance of El absorption below 100 MeV, means that
meson exchange currents do not make any appreciable contribution to
the total cross section - the El amplitudes are about an order-of-
magnitude larger than the spin-flip transitions that are influenced
by the magnetic moment operator. When these small amplitudes are
squared, as needed in calculating the total cross section, they give
negligible contributions.

The Ml amplitudes that appear in the expression for the differential cross section are of the spin-conserving type and are
therefore also uneffected by the magnetic moment operator. However,
the expression for the differential polarization has the form:

$$p \left(\frac{d\sigma}{d\Omega}\right) = A \sin \theta + B \sin \theta \cos \theta$$

where the first term represents El-Ml interference that contains Ml
amplitudes of the spin-flip variety, and these are changed by the
inclusion of meson effects.

Drooks et al.[36] have searched for meson effects in the $d(\gamma,\vec{n})p$
at photon energies between 6 and 15 MeV, and at angles close to 60°,
90°, and 120° by measuring the polarization of the photoneutrons
with ^{12}C, absolutely calibrated in a true neutron double-scattering

TABLE 3

THE POLARIZATION OF PHOTONEUTRONS FROM THE REACTION

$\gamma + d \rightarrow \vec{n} + p$

IN THE PHOTON ENERGY RANGE 6 TO 16 MeV. [36]

REACTION ANGLE (θ°_{LAB})	PHOTON ENERGY RANGE (MeV)	POLARIZATION	^{12}C ANALYZER ANGLE (θ°_{LAB})
60	6.32- 7.30	-0.121±0.035	50
60	8.21- 9.00	-0.099±0.040	50
60	10.58-11.37	-0.110±0.088	50
60	12.04-13.94	-0.103±0.071	50
60 Mean = 9.85		Mean = -0.111±0.024	
90	6.53- 7.58	-0.124±0.023	50
90	6.65- 6.91	-0.156±0.109	130
90	7.07- 7.66	-0.131±0.044	130
90	8.53- 8.97	-0.040±0.057	130
90	8.55- 9.40	-0.110±0.026	50
90	9.68-10.69	-0.084±0.042	130
90	11.10-12.03	-0.080±0.059	50
90	12.35-15.36	-0.135±0.044	50
90 Mean = 9.32		Mean = -0.112±0.013	
121.5	6.76- 7.88	-0.123±0.021	50
121.5	6.93- 7.17	-0.126±0.110	130
121.5	7.34- 7.96	-0.178±0.056	130
121.5	8.87- 9.37	-0.139±0.073	130
121.5	8.91- 9.83	-0.129±0.025	50
121.5	10.09-12.11	-0.110±0.044	130
121.5	11.67-12.94	-0.128±0.051	50
121.5	13.03-16.33	-0.139±0.041	50
121.5 Mean = 9.83		Mean = -0.129±0.013	

188

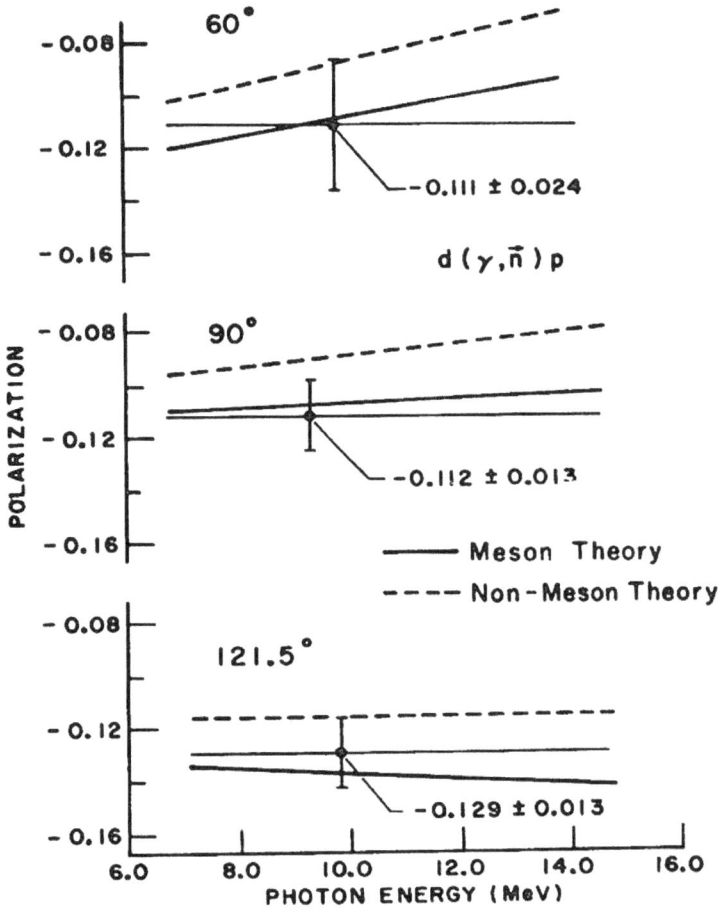

Fig. 9. The measured values of the polarizations of photo-neutrons from the reaction $d(\gamma,\vec{n})p$ (Drooks et al.[36]) compared with the non-meson and meson theories of Hadjimichael.[18]

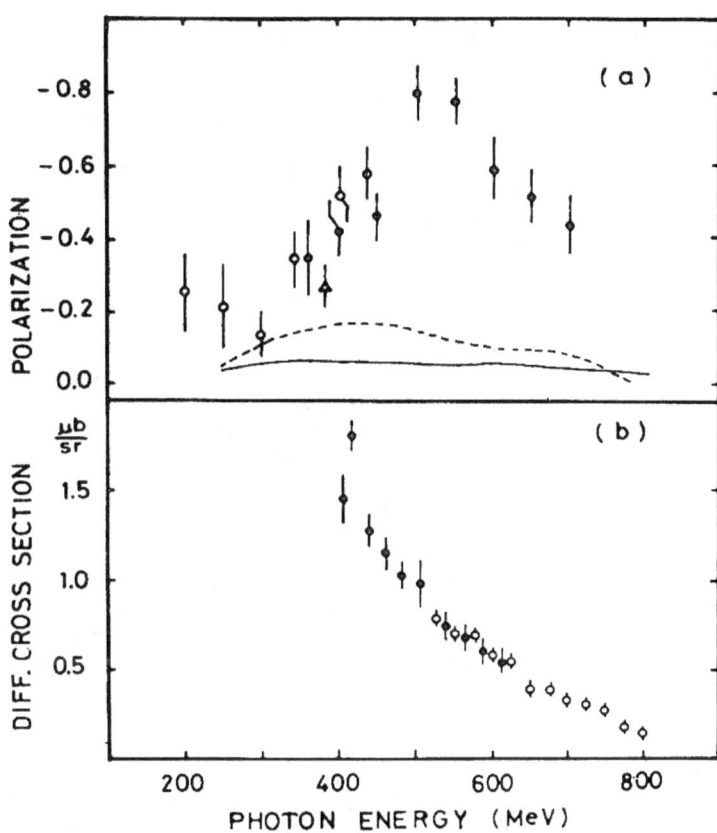

Fig. 10.   The values of the 90°-polarization of photoprotons
            from the reaction d(γ,p⃗)n obtained by Kamae et al.[22]
            [a], and the non-resonant differential cross section [b].

experiment.[38]   The target consisted of a cylinder of liquid
deuterium in a thin-walled cryostat.   The unpolarized photons were
generated by stopping a 32-MeV analyzed, pulsed beam of electrons
from the Yale LINAC in a thick bremsstrahlung target.   The results
are shown in Figure 9, and are tabulated in Table 3.   At all three
angles, the measured polarization is greater in magnitude than the
predictions of the classical result of Partovi; there is evidence
that the measurements support the calculation of Hadjimichael that
include meson exchange currents.   Indeed, at 90°, the measured value
of 0.112 ± 0.013 is almost 2 1/2 - standard deviations away from
the Partovi prediction, and is but 1/2 - a standard deviation from
the Hadjimichael prediction.

    Every effort was made during this long experiment to eliminate
sources of systematic errors.   (The final result was obtained from
five independent measurements, made over a period of 1 1/2 years
using different converters, beam stops, collimators etc.)   A full
scale Monte Carlo calculation was made to obtain the true "point"
polarizations.   It will certainly take a major effort to reduce the
present errors on the polarization from about 10% to less than 2%.
At first sight it would be better perhaps to make measurements at
higher energies, where larger effects are expected - but there are
no absolutely calibrated neutron polarization analyzers at energies
above 20 MeV, and, indeed, the intrusion of higher multipoles make
the calculations more difficult, and less certain.
    2.   Intermediate energies.
    Last year, Kamae et al.[22] reported the observation of a re-
markable resonance-like anomaly in the proton polarization from the
photo-disintegration of the deuteron, see Figure 10.   The "resonance" appears at a photon energy of 530 MeV and has a "width"
of about 200 MeV.   The proton polarization is sensitive to imaginary
interfering amplitudes.   The resonance-like structure has been interpreted as a new state of matter;[23] a dibaryon, Δ-Δ resonance(s-wave,
$T = 0$, $J = 3$).   This resonance gives an imaginary amplitude that
interferes with the (almost) real amplitudes (Born + NN).   The links
between this work and certain aspects of particle physics is clear -
the calculations need the $\Delta\Delta\pi$, $\Delta\Delta\eta$, $\Delta\Delta\rho$ and $\Delta\Delta\omega$ coupling constants.
It should be noted that there is no evidence of this interesting
new state in the differential cross section of the photoprotons (see
bottom of Figure 10).

IV CONCLUSIONS
    Nowhere in Nuclear Physics do we see clearer examples of the
role of meson currents than in the study of the interaction between
electromagnetic radiation and the two-nucleon system.   Surprises
constantly appear, and they continue to encourage experimentalists
to devise more sensitive tests of those important new branches of
theoretical physics that are bridging the gap between classical
Nuclear Physics and Particle Physics.

262   F. W. K. FIRK

REFERENCES

1. J. Chadwick and M. Goldhaber, Nature 134 (1935) 237
2. H. A. Bethe and R. E. Peierls, Proc. Roy. Soc. A148 (1935) 146
3. E. Fermi, Phys. Rev. 48 (1935) 570
4. G. Breit and E. U. Condon, Phys. Rev. 49 (1936) 904
5. W. S. Rarita and J. Schwinger, Phys. Rev. 59 (1941) 556
6. J. J. DeSwart, Physica 25 (1959) 233
7. W. Zernik, M. L. Rustigi and G. Breit, Phys. Rev. 114 (1959) 233
8. F. Partovi, Ann. of Phys. 27 (1964) 79
9. R. D. Nunemaker, Thesis, Yale Univ. (1968)
10. G. Breit, Proc. Inter. Conf. on Photonuclear Reactions and
Applications, Asilomar, Ca (1973)
11. M. LeBellac, F. M. Renard, and J. Tran Though-Van, Nuo. Cimento,
33 (1964) 594 and 34 (1964) 450
12. R. J. Adler, B. T. Chertok and H. C. Miller, Phys. Rev. 2C
(1970) 69
13. M. Gari and A. H. Huffman, Phys. Rev. C7 (1973) 994
14. D. -O. Riska and G. E. Brown, Phys. Lett. 388 (1972) 193
15. A. E. Cox, S. A. R. Wynchank and C. H. Collie, Nucl. Phys. 74
(1965) 481
16. H. P. Noyes, Nucl. Phys. 74 (1965) 508
17. F. Villars, Helv. Phys. Acta. 20 (1947) 476
18. E. Hadjimichael, Phys. Lett. 46B (1973) 147 and private communication
19. H. Arenhövel, W. Fabian and H. C. Miller, Phys. Lett. 52B (1974)
303
20. M. Gari and B. Sommer, Phys. Rev. Lett. 41 (1978) 22
21. J. S. Levinger and H. A. Bethe, Phys. Rev. 78 (2) (1950) 115
22. T. Kamae, I. Arai, T. Fujii, H. Ikeda, N. Kajiura, S. Kawabata,
N. Nakamura, K. Ogawa, T. Takeda and Y. Watase, Phys. Rev. Lett. 38
(1977) 468
23. T. Kamae and T. Fujita, Phys. Rev. Lett. 38 (1977) 471
24. D. Cokinos and E. Melkonian, Phys. Rev. C15 (1977) 1636
25. R. G. Arnold, B. E. Chertok, I. G. Schröder and J. L. Alberi,
Phys. Rev. C6 (1973) 1179
26. W. B. Dress, Claude Guet, Paul Perrin and P. D. Miller, Phys.
Rev. Lett. 34 752
27. D. E. Alburger, Phys. Rev. Lett. 35 (1975) 813
28. E. D. Earle, A. B. McDonald, O. Häusser and M. A. Lone, Phys.
Rev. Lett. 35 (1975) 908
29. E. D. Earle and A. B. McDonald, contribution to this conf.
30. B. Weissman and H. L. Schultz, Nucl. Phys. A174 (1971) 129
31. B. J. Hughes, A. Zieger, H. Wäffler and B. Ziegler, Nucl. Phys.
A267 (1976) 329
32. L. Allen, Phys. Rev. 98 (1955) 705
33. J. E. E. Baglin, R. W. Carr, E. J. Bentz and C. -P. Wu, Nucl.
Phys. A201 593
34. Y. M. Shin, J. A. Rawlins, W. Buss and A. O. Evwaraye, Nucl.
Phys. A154 (1970) 482
35. J. Ahrens, H. B. Eppler, H. Gimm, M. Kröning, P. Riehn,

H. Wäffler, A. Zieger and B. Ziegler, Phys. Lett. $\underline{52B}$ (1974) 49
36.   L. Drooks, Thesis, Yale Univ. (1976) and L. Drooks, F. W. K.
Firk, H. L. Schultz and R. J. Holt, B. A. P. S. $\underline{91}$ (1976) 534
37.   R. J. Holt, F. W. K. Firk, R. Nath and H. L. Schultz, Phys. Rev.
Lett. $\underline{28}$ (1972) 114

The Wigner Distribution at 50:

The Unlikely Beginning of

Random Matrix Theory

Frank W K Firk

Lecture given at Brown University on November 10, 2006

Abstract: 50 years have gone by since the late Professor Eugene Wigner (Nobel Laureate in Physics, 1963) surmised that the distribution of adjacent spacings between neutron-induced resonances in nuclei has the form

$$p(x)dx = (\pi/2) \cdot x \cdot \exp\{-\pi x^2/4\} dx,$$

where

$$x = \text{spacing/average spacing}.$$

His argument was based on the properties of matrices in which the elements are chosen randomly and independently from some probability distribution. Random Matrix Theory is now applied in many fields, including Chaos Theory, Quantum Chromo-Dynamics, String Theory and Quantum Gravity, mesoscopic systems in Condensed Matter, traffic flow patterns, communication networks, stock price fluctuations, and Number Theory. The pioneering experimental and theoretical developments of the subject that took place in the 1950's are discussed from a personal perspective.

Introduction

On November 1st and 2nd, 1956, a *"Conference on Neutron Physics by Time-of-Flight"* was held at Gatlinburg, Tennessee. At the time, I was a member of the Electron Accelerator Laboratory at Harwell, England, where I was using a high-resolution neutron time-of-flight spectrometer to measure the total cross sections of low-energy neutrons interacting with nuclei. A few months later, I received a copy of the conference proceedings that contained

Professor Wigner's now-famous surmise of the form of the spacing distribution of neutron resonances (of the same spin and parity) in nuclei.

The period from the mid-1930's to the late 1970's was the Golden Age of Neutron Physics; widespread interest in understanding the physics of the nucleus, coupled with the need for accurate data in the design of nuclear reactors and weapons, made the field of Neutron Physics of global importance in fundamental Physics, Technology, Economics, and Politics.

Neutron Resonances

In the mid-1950's, a discovery was made that turned out to have far-reaching consequences beyond anything that those of us working in the field could have imagined. For the first time, we were able to study the structure of the continuum in a many-body system, at high excitation energies. This unique situation came about as the result of the following facts:

◘ Neutrons, with kinetic energies of a few electron-volts, excite states in compound nuclei at energies ranging from about 5MeV to almost 10MeV – typical neutron binding energies. Schematically,

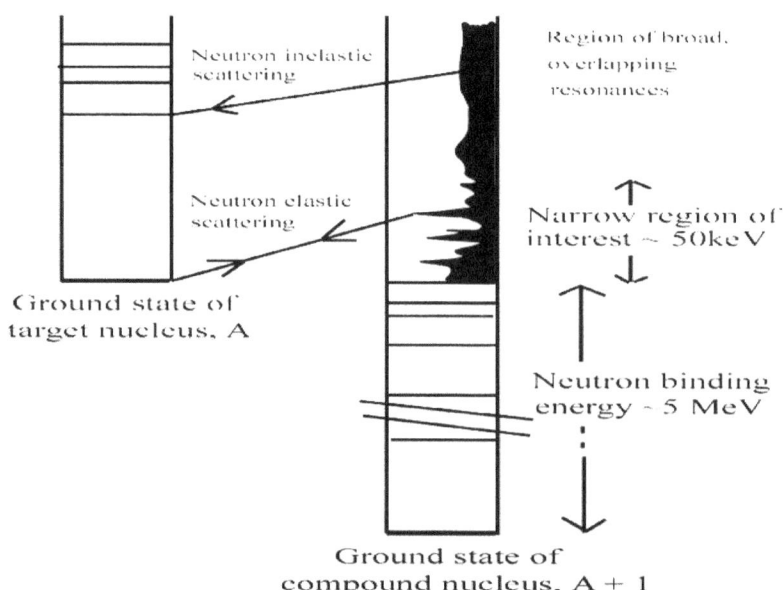

◻ The low-energy resonant states in heavy nuclei have lifetimes in the range 10^{-14} to 10^{-15} seconds, and therefore they have widths of about 1eV. The compound nucleus loses all memory of the way in which it is formed. It takes a relatively long time for sufficient energy to reside in a neutron before it is emitted. This is a statistical process. Typical average spacings of such resonances range from 10eV to 100eV.

◻ Just above the neutron binding energy, the angular momentum barrier restricts the possible range of values of total spin, J ($\mathbf{J} = \mathbf{I} + \mathbf{i} + \boldsymbol{\ell}$, where I is the spin of the target nucleus, i is the neutron spin, and ℓ is the relative orbital angular momentum).

◻ The neutron time-of flight method provides excellent energy resolution at energies up to several keV.

A 1-eV neutron travels 1 meter in 72.3 microseconds. At non-relativistic energies, the energy resolution ΔE at an energy E is simply:

$$\Delta E \approx 2E\Delta t/t_E,$$

where Δt is the *total* timing uncertainty, and t_E is the flight time for a neutron of energy E.

In 1956, my spectrometer had a timing uncertainty ≈ 200 nanoseconds for a 1-keV neutron. The flight path length was 56m. At 1keV, the resolution was $\Delta E \approx 3eV$.

The resolution ΔE depends on $E^{3/2}$.

In $^{238}U + n$, the excitation energy is about 5MeV; the effective resolution for a 1keV-neutron was therefore

$$\Delta E/E_{effective} \approx 6 \times 10^{-7}.$$

(At 1eV, the effective resolution was $\approx 10^{-11}$).

Two basic broadening effects limit the sensitivity of the method, they are:

1) Doppler broadening of the resonance profile due to the thermal motion of the target nuclei; it is characterized by the quantity $\delta \approx$

0.3√(E/A) (eV), where A is the mass number of the target. If E = 1keV and A = 200, δ ≈ 0.7eV, a value that may be ten times greater than the natural width of the resonance.

2) Resolution broadening of the observed profile due to the finite resolving power of the spectrometer.

In the early 1950's, the field of low-energy neutron resonance spectroscopy was dominated by research groups working at nuclear reactors. They were located at National Laboratories in the United States, the United Kingdom, Canada, and the former USSR. The energy spectrum of fission neutrons produced in a reactor is moderated in a hydrogenous material to generate an enhanced flux of low- energy neutrons. To carry out neutron time-of-flight spectroscopy, the continuous flux from the reactor is "chopped" using a massive steel rotor with fine slits through it. At the maximum attainable speed of rotation (about 20,000 rpm), and with slits a few thousandths-of-an-inch wide, it is possible to produce pulses with durations ≈ 1μsec. The chopped beams have rather low fluxes, and therefore the flight paths are limited in length to less than 50 meters. The resolution at 1keV is then ΔE ≈ 18eV, clearly not adequate for the study of resonance spacings ~ 10eV.

In late 1952, there were just four accelerator-based, low-energy neutron spectrometers operating in the world. They were at Columbia University in New York City (36″ cyclotron), Brookhaven National Laboratory (60″ cyclotron), the Atomic Energy Research Establishment, Harwell, England (15MeV traveling wave linear electron accelerator), and at Yale University (10MeV standing wave electron accelerator). The performances of these early accelerator-based spectrometers were comparable with those achieved at the reactor-based facilities. It was clear that the basic limitations of the neutron-chopper spectrometers had been reached, and therefore future developments in the field would require improvements in accelerator-based systems.

In 1955, I began building a new high-powered injector for the electron gun of the Harwell electron linear accelerator. My two colleagues in this project were John Gallagher and Gerry Reid, a member of the cyclotron group at Harwell. It took us about one year to construct an electron gun modulator that produced a high-current, high-voltage (> 70kV) pulse with a duration of 100 nanoseconds at 50kV – the voltage required to give the electrons the required injection velocity of 0.5c. The electron pulse could be injected at the instant of optimum RF power in the accelerator waveguides, thereby resulting in a very high peak accelerated current at maximum energy; we were operating in a new mode, called the "stored energy mode" of the accelerator. The pulse width had been reduced by a factor of ten, whereas the neutron yield decreased by only a factor of two. I was therefore able to construct a 56-meter flight path, thus achieving an energy resolution of 3eV at 1keV. During the years 1956-7, this was the highest resolution available, anywhere.

The 15-MeV Electron Linear Accelerator at Harwell (1952); from my original drawing.

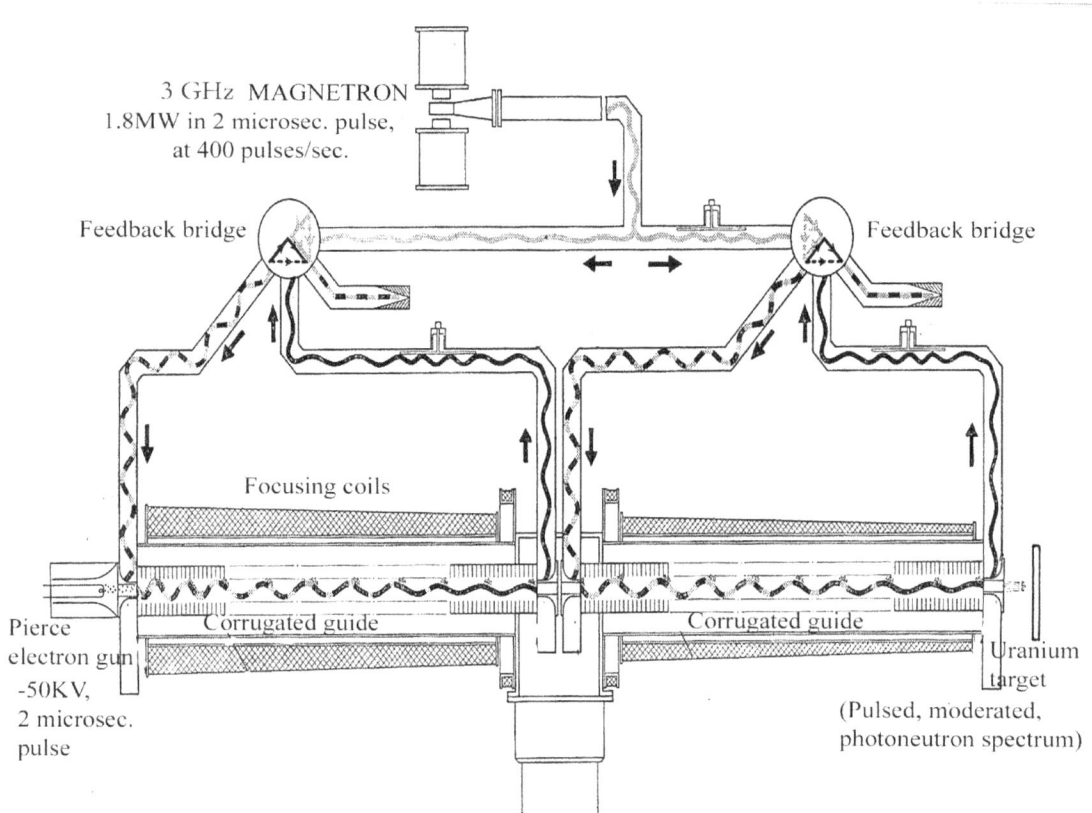

The RF feedback system of the Harwell accelerator

At the same time that we were developing the spectrometer at Harwell, Havens and Rainwater, and their colleagues at Columbia, were building a new 385-MeV proton synchrocyclotron a few miles north of the campus (at the Nevis Laboratory). The accelerator was designed to carry out experiments in meson physics and low-energy neutron physics. By 1958, they had produced a pulsed proton beam with a duration of 25 nanoseconds, and had built a 37-meter flight path. The hydrogenous moderator generated an effective pulse width of about 200 nanoseconds for 1keV-neutrons. By 1959, they had increased the length of the flight path to 200 meters, setting a new standard in neutron time-of-flight spectroscopy.

Nuclear Resonance Theory

The following brief review of nuclear resonance theory shows why it was important for us to study the spacing distribution of resonances in a range of nuclei.

The radial part of the Schroedinger equation for a spinless, s-wave neutron of mass m, scattered elastically from a square potential well of radius a is

$$- (\hbar^2/2m)\ddot{\varphi} + V\varphi = E\varphi$$

where $V = V_1$ for $r \le a$, $V = 0$ for $r > a$, and $\ddot{\varphi} = d^2\varphi/dr^2$.

The solutions are

$\varphi = A\sin(Ka)$ for $r \le a$,

and

$\varphi = B(e^{-ikr} - Ue^{ikr})$ for $r > a$.

A is a constant and B is chosen so that the incoming wave has unit flux

$B = (4\pi v)^{-1/2}$ where v is the neutron velocity.

U is the *COLLISION FUNCTION*; in terms of a phase shift δ it is

$U = e^{2i\delta}$.

The scattering cross section is

$\sigma_{nn} = (\pi/k^2)|1 - U|^2 = (4\pi/k^2)\sin^2\delta$.

At low energies, φ is almost a standing wave – "almost" because the neutron is unbound, and eventually leaves the well.

In the formal theory, a complete set of standing waves X_λ is constructed of which one closely resembles the true wave function, φ. In expanding φ in terms of X_λ, one term may dominate the expansion, *and this term then corresponds to the standing wave associated with the resonance.*

The X_λ's obey the same wave equation as φ:

$$- (\hbar^2/2m)\ddot{X}_\lambda + VX_\lambda = E_\lambda X_\lambda$$

with the boundary condition

$$r\dot{X}_\lambda = bX_\lambda I_{r=a}$$

where b is an arbitrary number.

The X_λ's form a complete orthogonal set, therefore

$$\varphi = \Sigma_\lambda A_\lambda X_\lambda$$

where

$$A_\lambda = \int_{0,a} X_\lambda \varphi dr.$$

Some algebra leads to

$$(\hbar^2/2ma)\{a\dot{\varphi}(a) - b\varphi(a)\}X_\lambda(a) = \int_{0,a} (E_\lambda - E)X_\lambda \varphi dr$$

and

$$\varphi = (\hbar^2/2ma)\Sigma_\lambda (E_\lambda - E)^{-1}X_\lambda^2(a)\{a\dot{\varphi}(a) - b\varphi(a)\}$$

or,

$$\varphi = R\{a\dot{\varphi}(a) - b\varphi(a)\} \text{ where }$$

$$R = \Sigma_\lambda \gamma_\lambda^2/(E_\lambda - E).$$

and

$$\gamma_\lambda^2 = (\hbar^2/2ma)X_\lambda^2(a).$$

The R-function is the principal quantity appearing in the formal theory of nuclear reactions.

The collision function, *U*, is related to R by considering the logarithmic derivative $(d\varphi/dr)/\varphi|_{r=a}$:

.

$$\varphi(a)/\varphi(a) = (1 + bR)/aR.$$

Now,

$$\varphi = I - UO,$$

we then obtain

$$U = e^{-2ika}\{(1 + bR + ikaR)/(1 + bR - ikaR)\}.$$

If there is a single resonance at E_0 then

$$R = \gamma_0^2/(E_0 - E),$$

and the cross section is

$$\sigma_{nn}^0 = (\pi/k^2)I2\sin(ka)e^{ika} - \Gamma_0/\{(E_0 - E + \Delta_0) - \Gamma_0/2\}I^2$$

where $\Gamma_0 = 2ka\gamma_0^2$ is the "width" of the resonance and $\Delta_0 = b\gamma_0^2$ is the "level shift". The simplest boundary condition is b = 0 – the standing wave resembles the true state as closely as possible. The cross section is then in the standard Breit-Wigner form.

The *total cross section* for an isolated s-wave neutron resonance, at low energies, can be written in the convenient notation

$$\sigma(x) = \sigma_0/(1 + x^2) + 2ka'\sigma_0 x/(1+x^2) + 4\pi a'^2$$

| Resonance term | Interference term | Potential scattering term |

in which

$$x = 2(E - E_{res})/\Gamma;$$

$\sigma_0 = 4\pi g \Gamma_n \cos(2ka')/k^2\Gamma$, the *peak* cross section when the wavenumber,

$k = k_{res}$ (on resonance));

$g = (2J + 1)/2(2I + 1)$, the spin weighting factor;

a' is the effective nuclear radius modified by the contributions of distant resonances, and

$\Gamma = \Gamma_n + \Gamma_\gamma$, the *total* width. ($\Gamma_n$ is the neutron width and Γ_γ is the total radiation width).

In an energy interval containing many resonances (of the same spin and parity), the average cross section can be calculated in terms of the individual resonance cross sections. For a single resonance,

$\int_{-\infty, +\infty} \sigma(E)dE = (\pi/2)\sigma_0\Gamma$,

therefore, in an interval containing many resonances with an average spacing <D>, we obtain

$<\sigma><D> = (\pi/2)<\sigma_0\Gamma>$

and

$<\sigma> = (2\pi^2 g/<k>^2)\{<\Gamma_n>/<D>\}$.

The key quantity $S = <\Gamma_n>/<D>$ is called the *strength function*.

It is the custom to normalize the neutron widths to 1eV, then

$S_{expt} = <\Gamma_n^0>/<D>$

where $\Gamma_n^0 = \Gamma_n\sqrt{(E/1)}$.

The study of strength functions involved the determination of average spacings, <D>, in different nuclei, and therefore we were led to study *spacing distributions*.

In the simplest potential model of the nucleus, the so-called "black nucleus model", the predicted value of S_{expt} is 1×10^{-4}, independent of mass number. The following diagram shows the measured values as a function of mass number. The essential features of this diagram were known by the mid-1950's. The fitted curve illustrates the sensitivity of the fit to the parameters of a deformed (complex) nuclear potential (an "optical model").

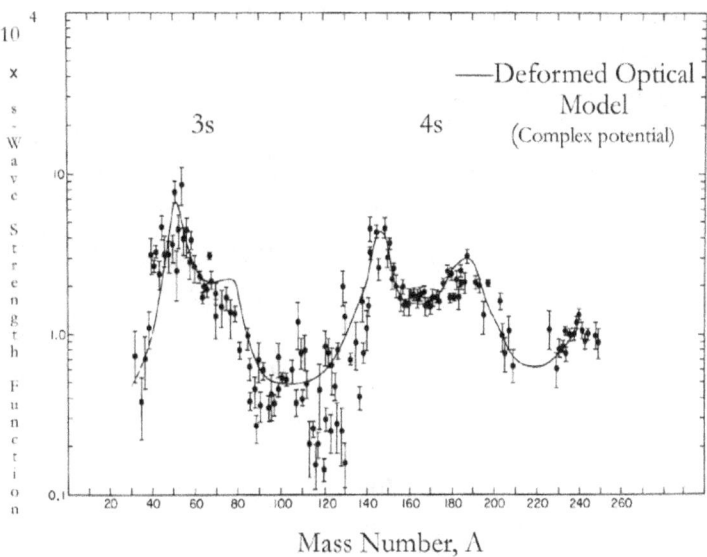

Mass Number, A

A deformed optical model fit to the observed s-wave strength function.

The following quotation is from Professor Wigner's presentation, 50 years ago; it is quintessential Wigner:

"Perhaps I am now too courageous when I try to guess the distribution of the distances between successive levels. I should re-emphasize that levels that have different J-values (total spin) are not connected with each other. They are entirely independent. So far, experimental data are available only on even-even elements. Theoretically, the situation is quite simple if one attacks the problem in a simple-minded fashion. The question is simply "what are the distances of the characteristic values of a symmetric matrix with random coefficients?"

We know that the chance that two such energy levels coincide is infinitely unlikely.

We consider a two-dimensional matrix

,

$$a_{11} \quad a_{12}$$

$$a_{21} \quad a_{22}$$

in which case the distance between two levels is $[(a_{11} - a_{22})^2 + 4a_{12}^2]^{1/2}$. This distance can be zero only if $a_{11} = a_{22}$ and $a_{12} = 0$.

The difference between the two energy levels is the distance of a point from the origin, the two coordinates of which are $(a_{11} - a_{22})$ and a_{12}. The probability that this distance is S is, for small values of S, always proportional to S itself because the volume element of the plane in polar coordinates contains the radius as a factor......

The probability of finding the next level at a distance S now becomes proportional to SdS.

Hence the simplest assumption will give the probability

$$(\pi/2)\, \rho^2\, exp\{ -(\pi/4)\rho^2 S^2 \}\, SdS$$

for a spacing between S and S + dS.

If we put $x = \rho S = S/{<}S{>}$, where ${<}S{>}$ is the mean spacing, then the probability distribution takes the standard form

$$p(x)dx = (\pi/2)\, x\, exp\{-\pi x^2/4\}dx,$$

where the coefficients are obtained by normalizing both the area and the mean to unity."

This is the now-famous Wigner distribution function. Wigner first introduced this form in 1951; however, he had not then taken the random matrix approach to the problem. In 1955, Landau and Smorodinsky, following Wigner's earlier work, deduced the same form.

It is interesting to note that the Wigner distribution is a special case of an important statistical distribution, named after Professor E H Waloddi Weibull (1887–1979), a Swedish engineer and statistician. The distribution was first published in 1939, and has been in widespread use since that time in statistical analyses in industries such as aerospace, automotive, electric power, nuclear power, communications, and life insurance. The distribution gives the lifetimes of objects and is therefore invaluable in studies of the failure rates of materials (including people!). The Weibull probability density function is

$$\text{Wei}(x; k, \lambda) = (k/\lambda)(x/\lambda)^{(k-1)}\exp\{-(x/\lambda)^k\}$$

where $x \geq 0$, $k > 0$ is the *shape* parameter, and $\lambda > 0$ is the *scale* parameter.

We see that Wei(x; 2, 2/√π) = p(x), the Wigner distribution.

Typical Weibull distributions are given in the following list

$$\text{Wei}(x; 1, 1) = \exp\{-x\}$$

$$\text{Wei}(x; 2, \lambda) = \text{Ray}(\lambda), \text{ the Rayleigh distribution,}$$

and

$$\text{Wei}(x; 3, 1) \approx \text{ the Normal distribution.}$$

The mean $= \lambda\Gamma(1 + (1/k))$,

the median $= \lambda\ln(2)^{1/k}$,

and

the mode $= \lambda(k - 1)^{1/k}/k^{1/k}$, if $k>1$.

As $k \to \infty$, the Weibull distribution has a sharp peak at λ.

Nuclear Physicists often mistakenly refer to the Weibull distribution as the Brody distribution (introduced in 1980).

In the 1920's, Professor Emil Gumbel, a mathematician at Heidelberg, introduced so called "extreme value distributions" in the theory of probability; these distributions are Weibull distributions.

In 1956, no more than 20 s- wave resonances had been clearly resolved in a single compound nucleus. Nonetheless, Don Hughes, Jack Harvey, and their collaborators, working at the fast-neutron-chopper-group at the high flux reactor at the Brookhaven National Laboratory, gathered their own limited data, and the data from all the neutron spectroscopy groups around the world, to obtain the first global spacing distribution of s-wave neutron resonances. Their combined results, published in 1958, showed a distinct lack of very closely spaced resonances, in agreement with the Wigner surmise.

By late 1959, the experimental situation had improved, greatly. At Columbia University, two students of Professors Havens and Rainwater (pioneers in the field of low-energy neutron time-of-

flight spectroscopy) completed their PhD theses; one, Joel Rosen, studied the first 55 resonances in ^{238}U + n up to 1keV, and the other, J Scott Desjardins, studied resonances in two silver isotopes (of different spin) in the same energy region. These were the first results from the new high-resolution neutron facility at the Nevis cyclotron. At Harwell, M C Moxon and I completed our work on the first 100 resonances in ^{238}U + n at energies up to 1.8keV; here are the observed resonances in the range 400 – 1800 eV.

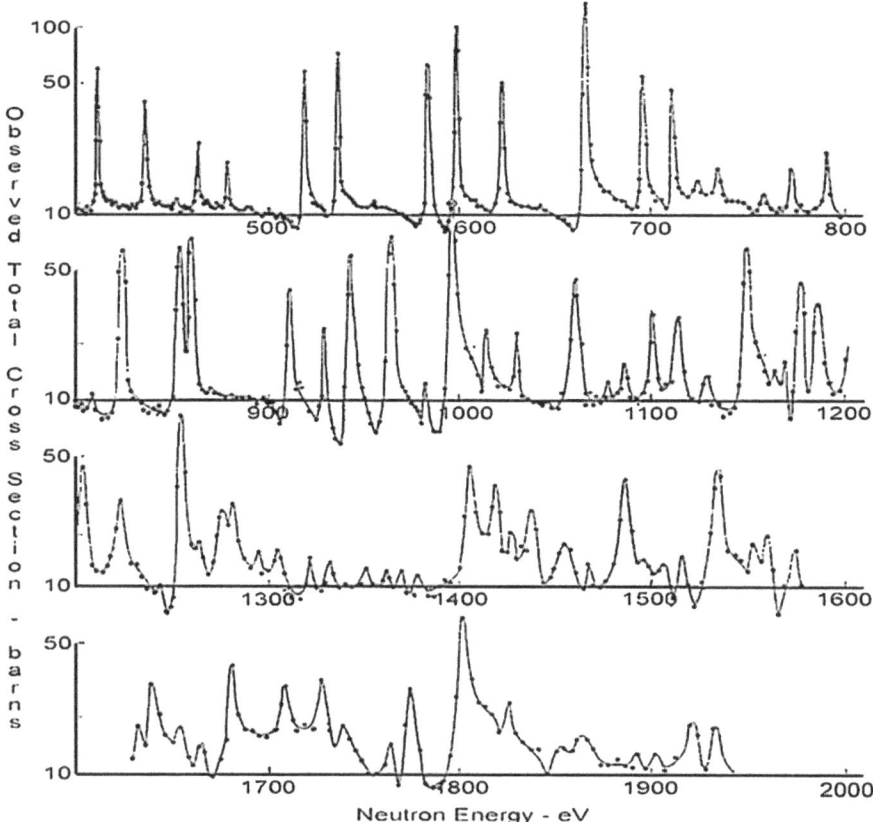

High resolution studies of the total neutron cross section of ^{238}U, (Firk, Lynn, and Moxon (1960)).

When we began this experiment in 1956, no clearly resolved resonances had been observed at energies above 500eV.

I presented these results at the Kingston International Conference on Nuclear Structure in late August, 1960 (see Firk, Lynn, and Moxon, (1960)). Moxon and I could not have analyzed our data without the invaluable theoretical contribution of my office-mate of many years, Eric Lynn, the author of the standard work in the field "The Theory of Neutron Resonance Reactions" (Oxford,

1968). The distribution of adjacent spacings of the first 100 resonances in the compound nucleus, ^{238}U + n, derived from our measurement, is shown. The result ruled out an exponential distribution and provided the best evidence, then available, in support of the proposed distribution.

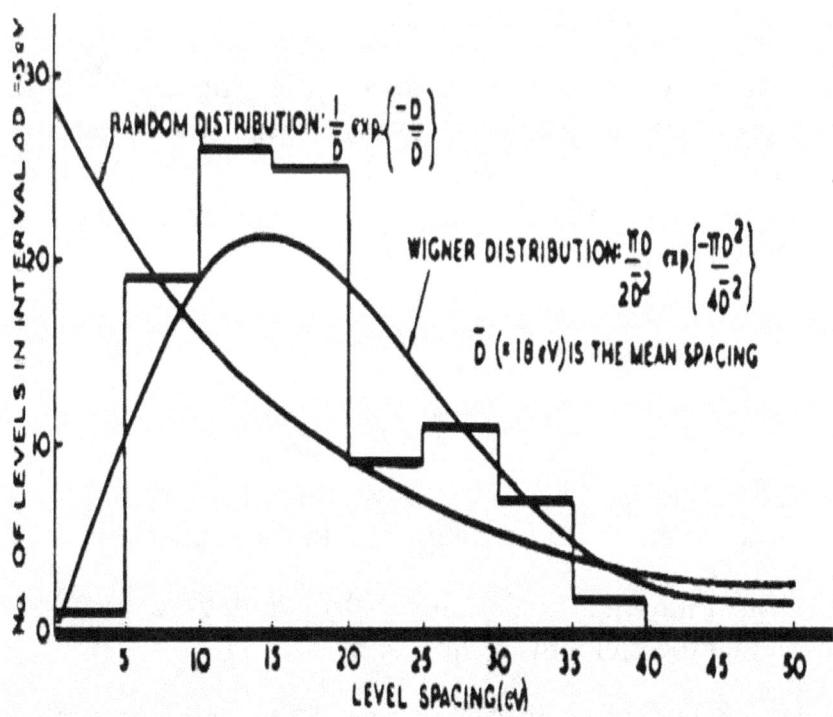

The observed spacing distribution of 100 resonances in ^{238}U + n.

208

Over the last half-century, further studies have not changed our basic findings. At the present time, almost 1000 s-wave neutron resonances in the compound nucleus ^{239}U have been observed in the energy range up to 20keV. The latest results, with their greatly improved statistics, continue to validate the form of the Wigner distribution.

An Introduction to Random Matrix Theory

A formal discussion of the theory follows: the many-body Hamiltonian matrix, H, (*infinitely* dimensional because there are infinitely many continuum states) of an excited heavy nucleus has an intractable form. H is therefore replaced by a collection (ensemble) of *finite* N x N (model) matrices in which the matrix elements are independently chosen from some probability distribution, p.

For a particular matrix, the statistical properties of its eigenvalues (corresponding to the energies of the compound nucleus) can, in principle, be calculated. An average over all matrices in the ensemble is taken, and the results obtained as N $\rightarrow \infty$.

The essential connection between the physical and mathematical aspects of the problem can be best illustrated by developing of the 2 x 2 case. We shall see that the physical constraints of rotational invariance and time-reversal invariance fix the form of p to be a Gaussian. Let

$$H = \begin{matrix} h_{11} & h_{12} \\ h_{21} & h_{22} \end{matrix},$$

a Hermitian matrix, invariant under rotations and time reversal.

The hermitian constraint means that $h_{12} = h_{21}$, and time reversal invariance means that the elements h_{ij} are real.

To reflect the statistical independence of the elements, the probability of finding a particular H (in which $h_{12} = h_{21}$) in an ensemble is

$$p(H) = p_{11}(h_{11}) \cdot p_{12}(h_{12}) \cdot p_{22}(h_{22})$$

The matrix H is defined in a basis, arbitrarily chosen; it is therefore assumed that the basis states, I1> and I2>, are invariant under an orthogonal transformation (the elements are real). The discussion will be limited to infinitesimal rotations (a finite rotation is generated from their sum). The transformed states are then

$$I1'> = \quad I1> + \varepsilon I2>$$

and

$$I2'> = -\varepsilon I1> + \quad I2>$$

In the transformed basis, the matrix elements are, to order ε,

$$h_{11}' = h_{11} + 2\varepsilon h_{12}$$

$$h_{12}' = h_{12} + \varepsilon(h_{22} - h_{11})$$

$$h_{22}' = h_{22} - 2\varepsilon h_{12}$$

Calculating the partial derivatives of $p(H)$ with respect to the three elements, we can relate $p(H')$ to $p(H)$; to order ε we obtain

$$p(H') = p(H) \{1 + \varepsilon[(2h_{12}/p_{11})(dp_{11}/dh_{11})$$
$$+ ((h_{22} - h_{11})/p_{12})(dp_{12}/dh_{12})(2h_{12}/p_{22})(dp_{22}/dh_{22})]\}$$

Invariance under a transformation of the basis states means that

$$p(H) = p(H')$$

This is true if the coefficient of ε vanishes. Therefore

$$(dp_{12}/dh_{12})(1/h_{12}p_{12}) = - (2/(h_{22} - h_{11}))$$

$$x [(1/p_{11})(dp_{11}/dh_{11}) - (1/ p_{22})(dp_{22}/(dh_{22})]$$

$$= \text{a constant, } - C, \text{ since the left hand side}$$

depends on h_{12} and the right hand side depends only on h_{11} and h_{22}. We then have

$$dp_{12}/p_{12} = - Ch_{12}dh_{12}$$

Integrating gives

$$p_{12} = (C/2\pi)^{1/2} \exp\{- Ch_{12}^2/2\}$$

The constant $\sqrt{(1/2\pi)}$ is determined by the chosen normalization. The constant C fixes the average magnitude of the matrix elements, and therefore determines the average level spacing.

The equations involving the h_{11} and h_{22} elements are

$$(2/p_{11})(dp_{11}/dh_{11})+Ch_{11}= (2/p_{22})(dp_{22}/dh_{22})+Ch_{22}= B$$

where B is a constant. We can choose B = 0 without loss of generality (B defines the zero of the energy scale).

Integrating gives

$$p_{11} = (C/4\pi)^{1/2} \exp\{-Ch_{11}^2/4\}$$

and

$$p_{22} = (C/4\pi)^{1/2} \exp\{-Ch_{22}^2/4\}$$

We note the Gaussian forms of the probability distributions, p_{ij}, of the matrix elements, h_{ij}.

The probability distribution, $p(H)$, is

$$p(H) = (4\sqrt{2})^{-1}(C/\pi)^{3/2}\exp\{-C(h_{11}^2 +2h_{12}^2 + h_{22}^2)/4\}$$

The eigenvalues of H are simply

$$E_\alpha = (h_{11} + h_{22})/2 + [(h_{11} - h_{22})^2 + 4h_{12}^2]^{1/2}/2$$

and

$$E_\beta = (h_{11} + h_{22})/2 - [(h_{11} - h_{22})^2 + 4h_{12}^2]^{1/2}/2$$
$$(E_\alpha > E_\beta)$$

Squaring these values, we obtain

$$E_\alpha^2 + E_\beta^2 = h_{11}^2 + 2h_{12}^2 + h_{22}^2,$$

and therefore

$$p(H) = \text{const. } \exp\{-C(E_\alpha{}^2 + E_\beta{}^2)/4\}$$

$$= \text{const. } \exp\{-C\text{Tr}(H^2)/4\}$$

It is now necessary to obtain the joint probability distribution in terms of the eigenvalues E_α and E_β, and an angle of rotation, θ, of the basis states.

The matrix H has been chosen to be real and symmetric therefore an orthogonal matrix, O, exists that will diagonalize H:

$$O^T H O = H^{\text{diag}} = \text{diag}(E_\alpha, E_\beta) \text{ where}$$

$$O = \begin{matrix} \cos\theta & \sin\theta \\ \\ -\sin\theta & \cos\theta \end{matrix}$$

the rotation matrix, used in its infinitesimal form to rotate the original basis states.

In the process of diagonalizing H, the $\{h_{11}, h_{12}, h_{22}\}$-space is transformed into the $\{E_\alpha, E_\beta, \theta\}$-space:

$$H = O H^{\text{diag}} O^T$$

Carrying out this transformation, and comparing matrix elements, we obtain

$$h_{11} = \cos^2(\theta)E_\alpha + \sin^2(\theta)E_\beta$$

$$h_{12} = \cos(\theta)\sin(\theta)(E_\alpha - E_\beta)$$

$$h_{22} = \sin^2(\theta)E_\alpha + \cos^2(\theta)E_\beta$$

The transformation means that the probability $p(h_{11}, h_{12}, h_{22})$, associated with the matrix elements, has become $p_{E,\theta}(E_\alpha, E_\beta, \theta)$, associated with the eigenvalues. To obtain the volume element in the new variables, it is necessary to calculate the Jacobian (a determinant)

$$|J| = |\partial(h_{11}, h_{12}, h_{22})/\partial(E_\alpha, E_\beta, \theta)|$$

$$= |\partial h_{11}/\partial E_\alpha, .. \text{ etc. }|$$

The partial derivatives with respect to θ are

$\partial h_{11}/\partial\theta = -2(E_\alpha - E_\beta)\sin\theta\cos\theta$

$\partial h_{12}/\partial\theta = \quad (E_\alpha - E_\beta)\cos(2\theta)$

$\partial h_{22}/\partial\theta = \quad 2(E_\alpha - E_\beta)\sin\theta\cos\theta$

All the partial derivatives with respect to E_α and E_β are functions of θ, alone. The three partial derivatives with respect to θ, in a single row of the determinant, have a common factor, $(E_\alpha - E_\beta)$.

We can therefore remove this factor from the determinant, and write

$\det(J) = (E_\alpha - E_\beta)f(\theta)$, where $E_\alpha > E_\beta$.

We therefore obtain

$p(h_{11}, h_{12}, h_{22})dh_{11}dh_{12}dh_{22} = (4\sqrt{2})^{-1}(C/\pi)^{3/2}$

$\times \exp\{-C(E_\alpha^2 + E_\beta^2)/4\}(E_\alpha - E_\beta)f(\theta)dE_\alpha dE_\beta d\theta$

Integrating over θ, leaves the joint probability distribution

$p_E(E_\alpha, E_\beta) = K(E_\alpha - E_\beta)\exp\{-C(E_\alpha^2 + E_\beta^2)/4\}$

K is a constant that can be calculated from the integral-over-angle.

The probability of a spacing S between two eigenvalues is obtained by integrating over E_α and E_β, thus

$p(S) = \iint p_E(E_\alpha, E_\beta)\delta(S - (E_\alpha - E_\beta))dE_\alpha dE_\beta$

(where the limits are $E_\beta < E_\alpha < \infty$ and $-\infty < E_\beta < \infty$)

$= (C/4)\cdot S\cdot\exp\{-C\cdot S^2/8\}$

The average spacing is

$$<S> = \int_0^\infty S\cdot p(S)dS = (2\pi/C)^{1/2}$$

so that

$C = 2\pi/<S>^2$

The correctly normalized distribution is therefore

$$p(S) = (\pi/2<S>^2) \cdot S \cdot \exp\{-\pi S^2/4<S>^2\}$$

If we introduce $x = S/<S>$, then

$$\mathbf{p(x)dx = (\pi/2) \cdot x \cdot \exp\{-\pi x^2/4\}dx};$$

this is the *Wigner probability distribution.*

The general case of a Gaussian Orthogonal Ensemble (GOE) of real, symmetric, n x n matrices can be developed from the 2 x 2 case, discussed above.

We need to transform the probability density distribution from the space of matrix elements to the space of the n-eigenvalues, and the υ-parameters that specify the orthogonal transformation matrix. Here, $\upsilon = n(n-1)/2$, the number of 2-dimensional subspaces in the n x n matrix; it is necessary to introduce υ because we carry out 2 x 2 infinitesimal rotations, in succession, in each subspace. In this way, the basis is transformed, and the matrix elements of H are found in the new basis (labeled h_{11}', h_{12}'...). The differentials dh_{11}, dh_{12}, ...dh_{nn} are thereby obtained. Invariance under a change of basis means that

$$p(H) = p(H')$$

In the general case, the forms of the probabilities, $p_{ij}(h_{ij})$, remain the same as found in the 2 x 2 case; they are determined by the symmetry conditions imposed on the hermitian matrices H. Each $p_{ij}(h_{ij})$ must have a Gaussian form. For example, the diagonal elements are found to be

$$p_{ii}(h_{ii}) = (C/4\pi)^{1/2}\exp\{-Ch_{ii}^2/4\},$$

and the non-diagonal elements are

$$p_{ij}(h_{ij}) = (C/2\pi)^{1/2}\exp\{-Ch_{ij}^2/2\}$$

The probability of a matrix H_{nxn} in the general ensemble is the product of the probabilities of its diagonal and non-diagonal elements; it is

$$p(H_{nxn}) = (2)^{-n/2}(C/2\pi)^{n(n+1)/4}\exp\{-(C/4)Tr(H^2)\}$$

214

We require the joint distribution of the eigenvalues, E_n.

The n x n matrices are diagonalized, and the distribution, $p(H_{nxn})$ is then transformed from the variables $(h_{11}, h_{12}, \ldots h_{nn})$ to new variables involving the eigenvalues E_α ($\alpha = 1$ to n) and the υ parameters θ_υ ($\upsilon = 1$ to $n(n-1)/2$).

The orthogonal transformation of the diagonal matrix H_{nxn}^{diag} to give H_{nxn} enables us to write each element h_{ij} in terms of the eigenvalues, E_α, and a complicated function of the variables, θ_υ. There are $\upsilon = n(n-1)/2$ rows of J that have eigenvalues in them, and

they are found to be linear in the eigenvalues. The Jacobian is therefore of the form

$$J \text{ (a determinant)} = F(\theta_1, \theta_2, \ldots \theta_\upsilon)\prod_{\alpha < \beta}(E_\alpha - E_\beta)$$

(a form that we obtained, in detail, in the 2 x 2 case).

Again, we assume that $p(H_{nxn})$ is independent of a rotation of the base states, and therefore

$$p(E_1, \ldots E_n; \theta_1, \ldots \theta_\upsilon)dh_{11}dh_{12}.. = F(\theta_1, \ldots \theta_n)\prod_{\alpha < \beta} |E_\alpha - E_\beta|$$

$$\mathbf{x} \exp\{(-C/4)\Sigma_\alpha E_\alpha^2\}dE_1..dE_nd\theta_1..d\theta_\upsilon$$

The first investigation of the distribution of successive eigenvalues for the above ensemble was carried out by Rozenzweig and Porter in the late 1950's, (work published in 1960). They numerically diagonalized a large number of matrices, with elements generated randomly, and constrained by the probability distribution, p(H). The analytical theory developed in parallel with their work: Mehta (1960), Mehta and Gaudin (1960), and Gaudin (1961). At the time, it was clear that, even with different assumptions concerning the chosen form of the probability distribution, the spacing distribution remained essentially the same. Remarkably, the n x n distributions had forms given, almost exactly, by the original Wigner 2 x 2 distribution.

The linear dependence of p(S) on the spacing S (for small S) is a direct consequence of the symmetries imposed on the Hamiltonian matrix, $H(h_{ij})$. Dyson, under the local influence of Wigner and Mehta at Princeton, became interested in the problem, and made a key contribution by showing that different results are obtained when different symmetries are assumed for H.

The impact of these developments was not immediate in Nuclear Physics. At the time, the main research endeavors were concerned with the structure of nuclei – experiments and theories connected with Shell-, Collective-, and Unified models, and with the nucleon-nucleon interaction. The study of Quantum Statistical Mechanics was far removed from the main stream.

Interestingly, the next development occurred in an area having nothing to do with Physics. In the field of Number Theory, perhaps the greatest unsolved problem has to do with the Riemann conjecture (that dates from the mid-19th century):

If $\zeta(z) = 1 + 1/2^z + 1/3^z + 1/4^z + \ldots 1/n^z + \ldots$ an infinite series, then the complex numbers z at which the zeta function has non-trivial zeros all have real part = ½. In 1914, Hardy proved that there are infinitely many zeros of the zeta function on the line x = ½. We do not know, to this day, if *all* of them are there.

In 1903, two giants of mathematics, Pólya and Hilbert conjectured (based on no evidence whatsoever!) that the non-trivial zeros of $\zeta(z)$ are of the Riemann form $½ + it_j$, where the numbers t_j are the eigenvalues of a naturally occurring Hermitian operator; they are therefore real. Hilbert used the phrase "spectrum" to describe the eigenvalues of the Hermitian operator, apparently by analogy with the optical spectra observed in atoms. This remarkable analogy pre-dated Heisenberg's Matrix Mechanics and the Hamiltonian formulation of Quantum Mechanics by more than a quarter-century. Not surprisingly, the Pólya-Hilbert line of argument was not pursued for almost 70 years. Around 1970, H L Montgomery, a

mathematician at the University of Michigan, investigated the relative spacing of the zeros of the zeta function. Let us recall that, if we have a series of points distributed randomly along a line, with average density normalized to 1, and we treat the coordinates of the points as independent random variables, then the probability of finding j points in a given interval of length x is the Poisson distribution

$$(x^j/j!)\exp\{-x\}$$

The eigenvalues of a random Hermitian matrix do not behave in this way; the probability of finding more than one eigenvalue in a short interval is less than given by the Poisson distribution; the eigenvalues "repel" each other. This effect is shown by the "pair correlation function". If t is an eigenvalue in the interval in which the relative spacing of the eigenvalues is normalized to unity, the correlation function gives the probability of finding the next eigenvalue near t + x. Montgomery investigated the pair correlation function for the zeros of the zeta function and he gave evidence that it has the asymptotic form

$$1 - [\sin^2(\pi x)/(\pi x)^2]$$

At a chance meeting between Montgomery and Dyson at Princeton in the early 70's, Montgomery showed his pair correlation function to Dyson, who recognized it as the pair correlation function of eigenvalues of random Hermitian matrices in a Gaussian Unitary Ensemble (an ensemble without time-reversal invariance). Dyson was working on the properties of this ensemble in connection with the development of the Wigner distribution of resonance spacings in nuclei. In a masterful numerical calculation of the distribution of spacings between zeros of the zeta function, Andrew Odlyzko tested the Montgomery conjecture by studying 70 million normalized zeros near 10^{20}. His computed correlation function shows remarkable agreement with the Montgomery form.

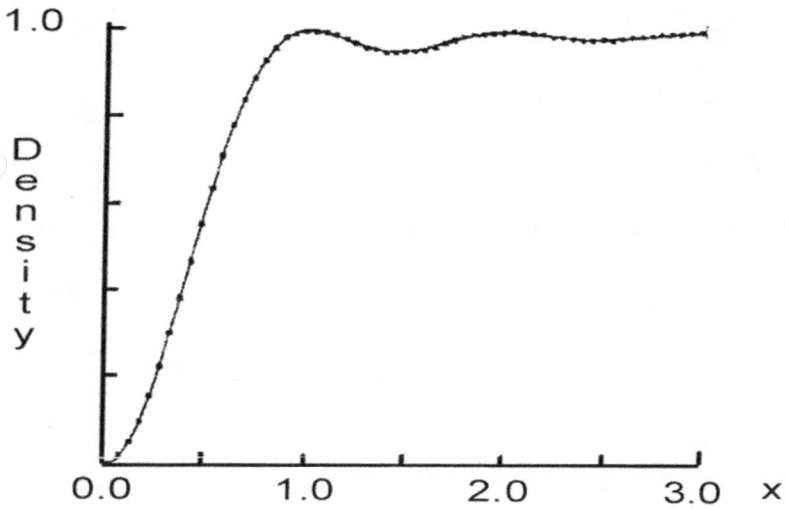

Odlyzko's test of the Montgomery conjecture, involving 70 million Riemann zeros near 10^{20}.

This work had a profound impact on developments in contemporary number Theory.

In the 1980's, Random Matrix Theory became a critically important tool in mesoscopic condensed matter systems, and in chaotic systems. My Yale colleague, Yoram Alhassid, published an exhaustive review of these developments in 2000. The theory, that has its origins in the narrow field of low-energy Neutron Resonance Spectroscopy 50 years ago, has become an essential part in many fields, ranging from Quantum Gravity to Communications.

References

The following books and articles provide the background material needed to gain a more complete understanding of the topics included in the present discussion:

1. Wigner, E. P., *Conference on Neutron Physics by Time-of-Flight*, Gatlinburg, TN Oak Ridge National Lab., ORNL-2309, p. 59, 1957

2. Harvey, J. A. and Hughes, D. J., *Phys. Rev.* **109**, 471, (1958)

3. Rosen, J. L., Desjardins, J. S., Rainwater, J., and Havens, W. W. Jr., *Phys. Rev.* **118**, 687, (1960)

4. Desjardins, J. S., Rosen, J. L., Rainwater, J., and Havens, W. W. Jr., *Phys. Rev.* **120**, 2214, (1960)

5. Firk, F. W. K., Lynn, J. E., and Moxon, M. C., in

Proc. Kingston International Conference on Nuclear Structure,

p. 757, University of Toronto Press, Toronto, 1960.

6. Mehta, M. L., *Nucl. Phys.* **18**, 395, (1960)

7. Porter, C. E., and Rosenzweig. N., *Phys. Rev.*, **120**, 1698, (1960)

8. Dyson. F. J., *J. Math. Phys.*, **3**, 140-175; 1191-1198;

1199-1215, (1962)

9. Porter, C. E., Editor, *Statistical Theories of Spectra*:

Fluctuations, Academic Press, New York, 1965

10. Mehta, Madan Lal, *Random Matrices, 3rd edition*,

Elsevier Inc., San Diego, CA, 2004

11. Lynn, J. E., *The Theory of Neutron Resonance Reactions*, Clarendon Press, Oxford, England, 1968

12. Firk, F. W. K., and Melkonian, E., *Total Neutron Cross Section Measurements*, in *Experimental Neutron Resonance Spectroscopy*,

Harvey, J. A., Editor, Academic Press, New York, 1970

13. D. J. Hughes, *Neutron Cross Sections*,

Pergamon Press, New York, 1957

14. Firk, F. W. K., *Neutron Time-of-Flight Spectrometers*,

in *Detectors in Nuclear Science*, D. A. Bromley, Editor,

North-Holland, Amsterdam, 1979

15. Alhassid, Y., *The Statistical Theory of Quantum Dots,*

in *Rev. Mod. Phys.*, **72**, 895, (2000)

References for articles reproduced here are:

1. "Total Neutron Cross Section Measurements",

   F. W. K. Firk and E. Melkonian *in*

   "Experimental Neutron Resonance Spectroscopy", ed. J. A. Harvey,

   Academic Press, New York and London, 1970

2. "Recent Developments in Neutron Detection", F. W. K. Firk *in*

   "Fast Neutron Physics Part II", eds. J. B. Marion and J. L. Fowler,

   Interscience Publishers, John Wiley & Sons, New York NY, 1963.

3. "Neutron and Photon Reaction Studies using Low-Energy

   Electron Linear Accelerators", F. W. K. Firk *in*

   "Third Accelerator Conference", Nuclear Instruments and Methods **28,**

   (1964) 205-219, North-Holland Publishing Co., Amsterdam

4. "Low-Energy Photonuclear Reactions", F. W. K. Firk *in*

   "Annual Review of Nuclear Science", ed. Emilio Segre, **20**, 39-78,

   Annual Reviews Inc. Palo Alto, CA, 1970.

5. "Neutron Polarization", F. W. K. Firk *in*

   "Proceedings of International Conference on the Interactions of Neutrons

   with Nuclei, Volume 1", Lowell, MA, ed. E. Sheldon, (1976) 389,

   USERDA Technical Information Center, Oak Ridge, TN.

6. "Neutron Time-of-Flight Spectrometers", F. W. K. Firk *in*

   "Detectors in Nuclear Science", ed. D. Allan Bromley,

   Nuclear Instruments and Methods, **162**, (1979) 529-563,

   North-Holland Publishing Co., Amsterdam.

7. "N-P Capture and the Photo-Disintegration of the Deuteron",

   F. W. K. Firk *in "Neutron Capture Gamma-Ray Spectroscopy"*,

   eds. R. E. Chrien and W. R. Kane, Plenum Press, New York, 1979.